HOME VIDEO

CHOOSING, MAINTAINING, AND REPAIRING YOUR HOME THEATER SYSTEM

Other books by Andrew Yoder

Pirate Radio Stations: Tuning in to Underground Broadcasts
Build Your Own Shortwave Antennas, 2nd Edition
Auto Audio: Choosing, Installing, Maintaining, and Repairing Car Stereo Systems
Shortwave Listening on the Road: The World Traveler's Guide
Pirate Radio: The Incredible Saga of America's Underground, Illegal Broadcasters
Pirate Radio Operations
The TAB Electronics Yellow Pages
The Complete Shortwave Listener's Handbook, 5th Edition
Home Audio: Choosing, Maintaining, and Repairing Your Audio System

HOME VIDEO

CHOOSING, MAINTAINING, AND REPAIRING YOUR HOME THEATER SYSTEM

Andrew Yoder

McGraw-Hill

New York San Francisco Washington, D.C. Auckland Bogota
Caracas Lisbon London Madrid Mexico City Milan
Montreal New Dehli San Juan Singapore
Sydney Tokyo Toronto

McGraw-Hill
A Division of The **McGraw-Hill** *Companies*

1 2 3 4 5 6 7 8 9 0 DOC/DOC 9 0 3 2 1 0 9 8

ISBN 0-07-076514-6 (HC)
ISBN 0-07-076513-8 (PBK)

The sponsoring editor for this book was Scott Grillo, the editing supervisor was Sally Glover, and the production supervisor was Tina Cameron. It was set in Windsor Light through the services of Cabinet Communications (Editing, Design, and Production).

Printed and bound by R.R. Donnelley & Sons Company.

This book is printed on recycled, acid-free paper containing a minimum of 50% recycled, de-inked fiber.

Contents

Acknowledgments

Because of the scope of this book, it could not have been written without help from many people and organizations. Thanks to Richard and Judy Yoder, Angie Piwonka, Dave Benson, Scott Hepler, Dave Homan, Gary Matthews, Aaron Bittner, Paul Sanders, David McCandless, John C. Baker, Stephen Moore, Brad Robison, and Lance Scott.

Thanks to Alasdair Patrick of Audioquest; Sam Fontaine of AudioControl; Ryne C. Allen of ESD Systems; Grant Lansdell, Kevin Gordon, and Karen Casper of JVC; Eric C. Kreis of Kenwood USA; Laurie Bianchi; Ralph Nichols of Polydax; Joanne Mustaphich of Harman/Kardon; Maria Horton of CD Technologies; Doyle Albee and Lynne Wall of Maxtor; Stacey Pierson of Voyetra Technologies; Jeffrey L. Hipps and Stephanie Rojas of Sherwood; Laurie Compton of Sony Electronics; David Berkus of Sanyo Fisher; Milan Hudecek of Rosetta Laboratories; Georgia Morgan of Drake; Joan McCarrell, Doan Hoff, and Marcia Walker of Yamaha; Justin Yamada of Morel Acoustics; and Wee Bee Audio.

Kevin Miller and Emilie Ann of Runco; Susie Capozza of IXMICRO; Amy Dattilo of Rotel; Rudolf Graf of North Country Radio; Rachel Baughn and John Bailey of *Monitoring Times; QST;* Debra Zimmerman of Princeton Graphics; Mina Shin of Samsung; Rotel; Michelle Houde of Esoteric Audio USA; THX;

MPact; Ursula Fleck-Jerwin and Sarah Hartley Edwards of Deutsche Welle; John Mark Ivey of Channel Master.

Thanks to April Nolan for giving the first book in this series a chance. Also to Scott Grillo for acquiring this book, Melanie Holscher and Sally Glover for the editing, and to Tina Cameron. Thanks to the rest of McGraw-Hill for getting this book out of the warehouse and onto peoples' bookshelves.

Very special thanks to Yvonne for improving my writing quality and for always pushing me to do better.

INTRODUCTION

Modern American culture is a peculiar amalgamation of anything that works well on television or movies. Sports and styles are popularized by what appears on TV. If you don't believe it, look at the success of NBA basketball after it was picked up by NBC or the increased popularity of NHL hockey once FOX began covering the games. The vast number of Jennifer Aniston "Rachel" haircuts attest to TV's fashion power.

What few people realize is that TV is on the verge of incredible changes and the medium will never be the same. You might be saying, "Yeah, yeah. I know. It's now a 'home theater' system, and you can listen to everything in stereo. Big deal." The techniques of improving the video quality and, in particular, the audio quality are very important, but these issues have very little to do with the upcoming video revolution.

As I wrote this book (and as you read this, so long as it's before 2006), the Federal Communications Commission (FCC) was planning to eliminate all standard analog TV broadcasting in the United States by 2006 and replace it with some form of DTV (digital TV). At this writing, there are 18 different digital broadcasting standards, so it's anyone's guess what system will actually be implemented. For more information concerning DTV and HDTV (high-definition television), see Chapter 9.

In the meantime, other forms of video are taking off and becoming successful. Direct broadcast satellite (DBS) broadcasting, a 1990s phenomena, has now attracted millions of subscribers in North America. Cable TV, the video wonder of the 1970s and 1980s, is the leading way to view television in the United States. And although the technology does not yet support broadcast TV quality, RealVideo and other forms of compressed digital video have begun to be broadcast via the Internet.

Today's television marketplace is a tug of war between the long-entrenched television networks, the cable TV suppliers and broadcasting networks, the DBS networks, the consumer equipment manufacturers, and the computer hardware and software manufacturers. At this point, the power is so evenly divided among the various interests that it's impossible to guess what the dominant form of television broadcasting will be by 2010--or even if there will be a dominant medium.

Television is the defining factor of video or home theater systems. The TV standards have always defined the equipment and standards that are used for video playback equipment. This, too, is changing with the release of DVD players and discs, which have a much greater video playback resolution than standard broadcast television. Now people are buying TVs, monitors, and video projectors with higher playback resolution and different screen dimensions, specifically for the purpose of playing back DVD.

As mentioned, no one really knows where video and television are headed as we cross into the 21st Century. This book, *Home Video,* was written to provide information on the current systems that are available, to look at the future possibilities, and (most importantly) to help you build the best system that you can. Right now. Also included is a chapter that provides some very light-duty troubleshooting, maintenance, and repair information to help you keep that system running properly.

Contact

I don't do equipment recommendations and I don't have the resources to respond to all of the questions about system designs, room acoustics, and equipment specifications. Read the audio/video magazines and newsletters for this information. However, if you would like to drop me a line concerning this book or your personal experiences with home theater systems, feel free to contact me:

Andrew Yoder
P.O. Box 642
Mont Alto, PA 17237 USA
ayoder@cvn.net

Because of the volume of mail that I've been receiving, I can't guarantee a response, but I will try.

HOME VIDEO

CHOOSING, MAINTAINING, AND REPAIRING YOUR HOME THEATER SYSTEM

1

Choosing a System

For decades, the television has been the center of the living room—a mighty appliance. Something to grab your attention while you're half-asleep and curled up on the couch. If you *really* wanted to be entertained, you went to a movie. That's the emotional difference between television and theater-shown movies: You can take a date to a theater, but you can't go home to watch TV. That's considered boring and cheap.

But technology and life are changing. With the proper equipment, design, and layout, you can install a home theater system that is superior to the theater experience. Best of all, you can avoid all of the theater pitfalls: sticky floors, women with big hair, and kids who laugh loudly at the wrong parts. Now that the technology is available, you just have to convince your date that you really are "going out" when you watch a movie at your home theater. But before you can start scheduling your weekends, you need to do some planning.

Definition

First of all, you need to define not just your requirements, but what home theater is. A true home theater system should follow the examples of real theaters. First of all, that means a video projector. No real theater uses a TV tube, so the projector is a must in a true system. The requirements for the audio system are a bit less exacting, in part because audio specifications vary significantly from theater to theater. Regardless, a low-noise hi-fi system is necessary. Finally, the environment should be somewhat like a theater, if you're going to call it a

Fig. 1-1 A nice, but casual, contemporary home theater design for the living room.

theater. Does it have theater seats or something a little more comfortable? Does it have red carpeting? Are the walls adorned with old movie posters? This is a true home theater.

But you certainly don't need to spend $50,000 to enjoy movies at home. And you don't need an antique brass popcorn maker to improve your system. *Home Video* is all about any type of home video, whether or not it is built into a full-out theater environment (Fig. 1-1).

RESEARCH

Most people don't bother to research prices; they just go to a store and choose whatever TV is on sale. But if you're serious about your home theater or your wallet, you'll have to do a bit of groundwork before you can complete your system. Possibly the toughest aspect of buying a video system is just searching through piles of magazines, catalogs, and sale fliers to pick out the components that you want. There's a mind-boggling array of audio/video equipment on the market and it can be rather difficult to research all of it.

The first method might not necessarily result in a mediocre or poor system, but with a little research, you should be able to put together a home theater that looks better, sounds better, is more flexible, or is less expensive than the typical TV system that people have lived with for decades.

If you love movies and TV enough to want your system to look and sound terrific, not just good, you'll probably try to buy the best system you can afford. Expect equipment manufacturers and dealers to take advantage of that desire.

BUYING NEW EQUIPMENT

Just about everywhere you look, someone is selling some sort of new home theater equipment. Getting a good system for a good price might require a little work, but it is entirely possible.

FOLLOWING THE MAGAZINES

The audio magazines on the market cover everything from consumer looks at mass-produced equipment to home brewing kits to deep technical reviews of ultra-expensive limited-production equipment. The home theater magazines are quite a bit different. Instead of digging into the technical or how-to aspects of audio/video, most of them are total color, glossy consumer magazines with large photographs of expensive systems in show homes. In fact, most seem to emphasize home and decor (or at least wildly outlandish home design) more than actually improving your system.

When it comes down to solid, basic home theater design, you can garner a few good tips and pointers from the home theater magazines, but you'll probably need to investigate other sources as well. Find what you can from the home theater magazines about video equipment, new products, and theories of home theater design, but read the stereo magazines for information concerning the audio side of home theater (Fig. 1-2).

Fig. 1-2 Some of the major audio periodicals in North America.

Consumer Reports and the *Consumer Reports Audio/Video Buying Guide* can also help with your purchases. This magazine rates a number of the readily available audio/video components. You won't find any of the high-end audio equipment or video projectors in the guide, but it does include the best of the consumer models. Of course, look carefully at the *Consumer Reports* ratings—some of what they consider to be important isn't terribly valuable to home theater enthusiasts. For example, when rating CD players, there is no rating for the sound quality—even though CD players are known to sound at least a bit different.

Because the audio, home theater, and consumer ratings magazines all have a different focal point, it's best to read as much as you can and determine what matches your own desires and capabilities. Make a checklist and get ready!

THE DIFFERENCES BETWEEN BUYING VIDEO AND AUDIO EQUIPMENT

As mentioned throughout this book, we are at a technological crossroads with television. The ripe, old analog system is about to pass into the world of digital TV, with vastly improved standards. This will completely change the way we look at the medium and the technology. The current technology also affects the way that we purchase equipment. We have one video broadcast standard and that is what TVs are manufactured to attain. The problem is that it is very easy for manufactures to reach the limits of video quality with the old standards. Thus, for decades, the biggest selling points of TVs have been price and features, not performance. This is much different from that of the audio market, where CDs and records are very accurate, and only very excellent equipment can reveal most of the intricacies of the music.

As a result, TVs are appliances and audio is an art. If you find a TV at a good price that has a decent-looking display, buy it. That's about all there is to it. For the most part, the same holds true for VCRs, although they have more variable characteristics—especially those with late 1990s innovations. When purchasing and installing audio equipment, you should be very careful to get exactly what you want.

In the upcoming years, as the standards change and new video technologies are developed, there should be much greater differences between expensive and cheaper equipment.

When shopping for a home theater system, remember that it is much more difficult to find an excellent audio system than it is to find a TV or VCR that looks about as good as you can get. Also, large TVs are very large, heavy, and fragile; they are very expensive to ship and most companies don't like to send them out. For the most part, because of the number of stores that sell TVs, buying a TV through the mail is unnecessary. So, the following sections are primarily dedicated to shopping for the audio equipment in your home theater system, although the comments loosely apply to video equipment as well.

BUYING FROM THE DISCOUNT DEPARTMENT STORES

If you get tired of running around, looking at ads, and researching, you might check out the stereo equipment at the local discount department store. After all, your "significant other" wants to try on some clothes and you want to look at the tennis racquets. There, you see equipment at great prices—complete minisystems for less than $149.95. Portable stereo systems (often called "boxes" around this part of Pennsylvania) are selling there for anywhere from $29.95 to $199.95. If you are lucky, they might even be selling some component stereo systems.

If you have listened to the television advertisements from the discount department stores, you will know that the reason why their products are so inexpensive is because they have the "buying power" of thousands of stores—they don't just buy by the crate or even by the truckload!

All of these claims are correct; these stores buy in volume and they do pass the savings on to you. However, they only have enough room for a small quantity of electronics equipment, and you won't find much of a selection. As a result, plenty of equipment is vying for space in this department store. What sells in the store is not based on the quality, but on the number of units sold per month. This, in turn, is not based entirely on quality, but also, on the price for the perceived number of features, size, and looks. The final judgment is decided by consumers (most of whom have not researched the market), not by audio professionals or hobbyists.

In general, the discount department stores attempt to give the consumer every feature possible for the least amount of money. This attitude is what sells products. You will see that the autoreverse cassette deck with Dolby C is much less expensive than the autoreverse cassette deck with Dolby C that you find in the stereo magazine advertisements. What you won't see is that the company that made the less-expensive deck probably skimped on the parts inside. Perhaps they saved a few cents per deck by using a less-sturdy PC board that will crack after a few years of use. Or maybe they used cheaper switches and knobs on the front panel that are much more likely to break. They almost certainly used cheaper heads, which means that the sound won't be as good and the heads will wear out quicker. The quality of the parts that go into a unit are much more important than the extra features and conveniences that you see on a deck. No where is this more apparent than when comparing the equipment in discount department stores to that of higher-quality consumer-grade equipment.

Of course, there is a time and a place for everything. If you are on an extremely tight budget and you aren't serious about putting together the best home theater, you might want to try this route. On the other hand, do not substitute cheap equipment for units that are well built if you plan to use the system for any length of time. Whereas the discount specials will reliably operate for maybe a few years, you can easily expect 5 or 10 years of good service from the better models. In the long run, researching the market and spending a few extra dollars on a solid system will be less expensive than buying inadequate systems that frequently break down.

One exception to the "cheap-equipment-at-national-chainstores" rule is Radio Shack. Although Radio Shack isn't a discount department store, it is basically a discount electronics department store. Because of its national-bargain nature, Radio Shack doesn't dig heavily into the home theater scene. They do, however, carry everything from low-grade to mid-grade consumer equipment, selling under the Optimus name. It can be rather tough to find comparison information on Radio Shack equipment in the stereo magazines. Evidently, Radio Shack has enough business from their own stores throughout the country and from the distribution of their sale fliers that they don't need to advertise in these magazines. In turn, the name Radio Shack is rarely mentioned in any of them. If you want to compare systems, you will have to look through *Consumer Reports* evaluations or check out the stereo section at the local Radio Shack store yourself.

MAIL-ORDER COMPANIES

If you have done your background work, you will find your best price on any given piece of audio/video equipment from one of the mail-order companies. Because these companies employ telemarketers, have computerized inventories, and sell thousands of units, they can afford to buy in huge quantities and, in turn, sell their equipment at a lower price than anyone else. Of course, this is just a generalization; some mail-order companies aren't at all inexpensive. Still, if you play the part of a smart consumer and shop around, you will probably be satisfied with the price (Fig. 1-3).

Fig. 1-3 ElectroWeb, one of the many companies that deal in video equipment via mail order.

Best of all, many of the high-end audio mail-order companies specialize in used equipment. You can't afford the new Pioneer Elite A/V receiver, the Conrad-Johnson power amplifier, the Toshiba DVD player, and the Thiel speakers, but you still have some money to play with? Start contacting the stereo dealers that also sell used equipment, you will find many components selling for as little as half of the typical asking price. The key to purchasing used equipment from dealers is whether the equipment is covered under warranty. The advantage of buying used equipment from a dealership, rather than from other audio enthusiasts (see the next section) is that the equipment is checked over and tested. If anything has failed, it will be replaced. In that respect, it is almost as safe as buying new equipment, yet you might save a considerable amount of money. The only equipment that you would expect to significantly degrade over five years or less are cassette decks and VCRs. These tape player/recorders have numerous moving parts and the heads wear down with use. A reputable used-equipment dealer should have a warranty that lasts at least 90 days. If the money-back guarantee period lasts for only a week or so, avoid the dealer.

You can locate a number of large regional audio dealerships through different audio magazines and via the Internet. Both *Audio* and *Stereophile* magazines feature classified ads and a number of dealer ads. *Audio* even features a "Dealer Directory" in the back so that you can locate some of the larger mail-order stereo shops. *Stereophile*, on the other hand, prints a nice, big list of audio stores where you can buy *Stereophile*. The list only consists of cities and telephone numbers, but with a little bit of searching, you can locate a number of high-end audio stores. Unfortunately, many of the stores on the *Stereophile* list are retail only—no mail order. However, *Videomaker* and *Video Magazine* are filled with plenty of ads for video equipment and even a few pieces of audio equipment.

Dig through these advertisements and make a list of phone numbers and addresses. Be sure to ask the salesperson if the store sells via mail order or if it is retail only. If you are only interested in a few particular models, call each company for price information. Always ask if they sell used, demo, or closeout equipment; you might find some excellent deals that aren't listed in the company's recent ads or catalogs. Compare the prices and go for the best deal. Unfortunately, most of these stores focus on stereo equipment. Some also sell new video equipment, but few sell used models.

In my opinion, it is best to make a list with a number of different components that would be satisfactory for your needs. Then, pick up the company's catalog or, if they don't have a catalog, ask the customer service representative about some of the recent sales. That way, if some mail-order company has a magnificent sale or closeout on a system that might be a little better than you thought you could afford, you'll have that extra option to consider.

Unfortunately, many mail-order companies don't have customer service representatives that are abreast of the current catalogs, like traditional showroom sales people are. To make matters worse, it can sometimes be difficult to talk to the customer service representatives for more than a few minutes. You're usually lucky if you can get out more than your name, address, credit card number, and catalog order number. To get a feel for what mail-order companies you can and can't talk to, call around, work with the companies that you like best, and write them a supportive letter or e-mail to let them know that you liked their customer service policy.

Of course, to order from one of the mail-order companies, you either need to have a credit card out and ready or you have to send a check or money order to the company's address. Either way, the concept of sending hundreds or even thousands of dollars off to some address that you saw in a magazine isn't terribly appealing. What if that one-million square-foot warehouse that is packed with stereo equipment on the waterfront is actually just an empty apartment in Brooklyn with two telephones? You really have no way of knowing, unless you have visited the store's showroom or unless you know someone who has ordered from that company before. Although you can't be certain of the integrity of the company that you are ordering from, one good method to help you evaluate a company's performance is to check a few month's worth of magazines for advertisements. If you can find that a company has been in business awhile, it's a good sign. It's an even better sign if you can find advertisements from that particular company that are at least several years old—and the ads in question have gotten larger in size. That probably means that the mail-order company has performed well over the years and has, in turn, consistently built up a larger customer base.

Still, even after researching the mail-order company, you might have reservations about sending that much money. That's a natural reaction, and it is always better to be cautious than to risk your money. So, if you have found a company that you would like to order from, only order part of what you want—maybe only a relatively inexpensive set of speakers. That way, if the company sends everything quickly and you are satisfied with their service, you can order the more expensive components (such as amplifiers, CD players, speakers, etc.) later. If you are not satisfied with the service, write a letter to the company and to the Better Business Bureau, if necessary. Chances are that you will be very happy with the service that you receive from a mail-order company.

BUYING USED EQUIPMENT FROM OTHER SOURCES

If you are daring and like to live life on the edge, then you might be game for home theater Russian roulette. Except, in this game, you win if you hit a hot component and get reamed monetarily if you pull a dud. Or, you might be handy with a soldering iron and some test equipment and be perfectly capable of repairing your own equipment and saving a fortune. Either way, buying used equipment from somewhere other than a mail-order company is not for the faint of heart.

As you might expect (especially if you frequent yard sales, auctions, or flea markets), you can find nearly any type of equipment in any condition for virtually any price. It's only a matter of looking to find what you want or need. Unfortunately, you probably will have to sift through hundreds of pieces of equipment until you hit the jackpot. Still, with patience and determination, buying used can be both rewarding and fulfilling.

Large (27 inches and larger), good-quality used televisions should start flooding the market at low prices. The high-tech home theater buffs will dump their TVs as soon as the HDTV receivers and the upwardly/downwardly compatible receivers are introduced. Then the supply of used NTSC (National Television Standards Committee) TVs will balloon even further above the current demand, which isn't exactly soaring.

Two more types of home theater equipment are now becoming easy and inexpensive to purchase used: laser disc players and mono VCRs. Laser disc players began to dent the sales of VCRs, but before they could really catch on, the DVD was introduced, with plenty of advantages over the laser disc. Video enthusiasts are already beginning to dump their laser disc players and video CDs for the new DVD technology. Considering the number of available movies in laser disc format, you should have no problem finding lots of used movies for sale in the newspaper classifieds or on the Internet.

Used mono VCRs are quite a bit less desirable, but quite a bit less expensive as a result. Even in the early to mid-1990s, stereo VCRs cost hundreds of dollars, and it was technology for the serious home theater enthusiast. But now, four-head stereo VCRs are ringing up at less than $200 (U.S.), not much more than the cost of a decent cassette deck. With such low prices for stereo VCRs, the demand for a used mono VCR is minuscule. Considering the wear on moving parts and on the heads, used VCRs depreciate significantly as it is. Because audio is an important element (not just a novelty) in the reality of the home theater experience, go for a used mono VCR only if your system is following a very tight budget (Fig. 1-4).

Fig. 1-4 Here's a near-ultimate in thrifty viewing. On the bottom is an older mono VCR. On top is an ancient Commodore computer monitor that was rescued from the junk heap and returned to service.

WHERE TO BUY USED EQUIPMENT

If you are a "slave to convenience," you will quickly tire of the used equipment game and pick up your home theater components at premium prices in a rental store. This route is fine if you have thousands of dollars ready for a system—at least you won't have to take any risks or have many hassles this way.

CLASSIFIED ADS

One easy way to find the type of equipment that you want, however, is to check out the classified advertisements in the local newspapers, in the shopper-type free classified papers, or on the Internet. It's much easier to sift through classified ads than it is to search through thousands of pieces of equipment at a flea market, electronics show, or auction.

The classified ads are excellent because they're a clearinghouse for audio/videophiles and technoids who constantly buy new equipment and sell off their old components. The prices in the classified ads are often very good—usually about 1/2 to 3/4 of the price that they originally paid for the equipment. However, they are not quite as inexpensive as those at flea markets. Most people who will take the time and money to put a classified ad in a daily newspaper will expect to receive a price that is closer to what they originally paid for the components.

When scouring the classified ads for good prices, keep abreast of the discount prices from the stereo distributors. What sounds like a great deal might be a dud. The audiophiles and technoids commonly buy a piece of equipment with new features as soon as it hits the market. They wind up paying the premium prices to be the first person around to have whatever it is that they have. When something new rolls around, they will buy it and sell off the old equipment. Even if they are selling the particular piece of equipment for 66% of what they paid for it, you might be able to find a stereo discount store that has a closeout on the same model, but is selling it for less than half of the original list price.

Another trap to beware of is people who purchase new equipment, raise the price, and resell it in the classified ads as new or almost new discounted equipment. This practice isn't common, but occasionally someone will try to make a few bucks off of an unknowledgeable person in this manner.

When you find a component that you are interested in buying, make a list of questions to ask the owner. In addition to the type-specific questions that you need to write down, ask:

* When did you buy it?
* Where did you buy it?
* Did you buy it new or used?
* Why are you selling it?
* Have you ever had any problems with it?
* Do you have the manuals and the manufacturer information?

Of course, not everyone will be honest with you, but some people will be. A few signs to beware of are physically damaged or dirty equipment. Physical perfection is not a must, but heavy wear can cause problems in some equipment. For example, an amplifier can usually get scratched and worn, yet still work perfectly. But a worn VCR might have more problems—heads can wear out; fragile buttons can wear out or become intermittent as a result of dust, smoke, or grease; and motors and gears can break down and strip. If you know when the equipment was manufactured, you can figure out whether or not the owner is telling the truth about buying the unit new. Also, when checking the physical condition of the equipment, be sure to check the serial numbers or equipment part numbers. Also beware of equipment that has the identification stickers ripped off, and do not buy any components that have the numbers scratched or etched out. This stuff is almost certainly stolen. Buying stolen equipment encourages thieves, and it could get you in trouble as well.

The best way to find out whether or not a component works is to test it. Few people will bother to try to sell an unworking system if you have a chance to test it—it's easy see if it works and how well. If the component that you are checking is anything but a speaker, check to see if the power indicator (usually an LED) lights. Then, be sure that the wires are all connected. Otherwise, in the case of speakers or amplifiers, the owner might let you listen to the system, including the component that you want to buy. Check to be sure that the particular amp or speaker system is actually connected to the system that you're listening to.

If you're checking out a VCR or A/V receiver, be sure to press all of the buttons and make sure that they work. Either a particular control system in the unit could be knocked out or a button could be broken or intermittent. You won't know if you don't check them. Don't just let the owner press a few buttons and say, "See, it works great--are you going to buy it?" Hastiness and high-pressure sales are warning signs that the system doesn't work as well as it should, that the price is too high, or that the owner is an impatient jerk. Beware in any case because if the owner is a jerk, he or she won't be reliable and won't help out if you run into any problems.

After all of these negative signs to avoid and potential reasons to distrust the owner, there is one very positive sign to be on the look out for. It is an excellent sign if the owner has the original box, manuals, installation information, and/or receipts. For one thing, it rules out the possibility that the unit was stolen. Also, the component is less likely to have been bought second hand by the original owner. One of the most important positive signs about having all of this information is that if the owner was so careful with the information that pertained to the component, he or she was probably just as careful (maybe more so) with the component itself.

If you are interested in buying a system from someone on the Internet (on one of the Usenet groups or on the classified section of an audio Web page), beware. There is no way for you to know if the seller is telling the truth or if they will take your money and disappear. On your side are the other users on the group, who will often alert everyone if someone with a bad reputation is offering something for sale. There are still some rip-offs via computer systems, but there are also a number of bargains (Fig. 1-5). The only way for the system to work well is for the users to be honest and self-policing. If you plan to purchase a piece of equipment via this method, it is best to buy from someone who has had a lengthy presence on the system and has successfully sold

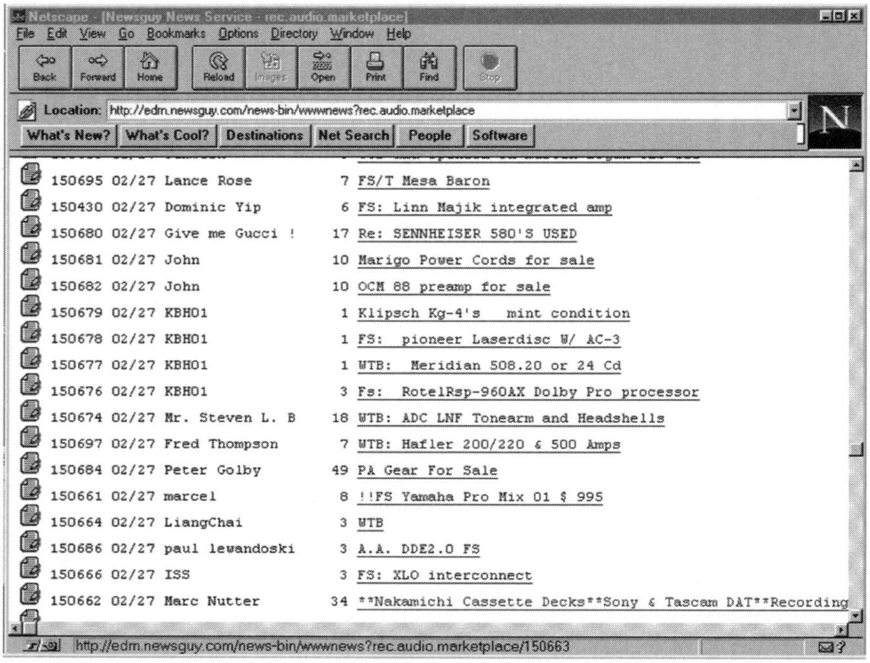

Fig. 1-5 A very small sampling of the classified ads on the rec.audio.marketplace Usenet group.

items in the past. Also, it's best not to put your life savings on a whole system that you see advertised. Order only as much as you can afford to lose. If it works, you've got a great deal; if not, at least you won't have to sell your car to cover the rent or a mortgage.

A variation of the Internet classifieds theme is E-Bay's Auction Web. This is a giant on-line auction with thousands of items, buyers, and sellers. The concepts behind Auction Web are simple: The seller pays E-Bay a small fee to list an item on their Web page. The item is listed for seven days. In that time, a buyer (who must be registered, free of charge, with E-Bay) can bid on the item. Then, the Web page lists the high bidder and the latest going price. At the end of the auction, the buyer and seller e-mail each other to work out the payment and shipping arrangements. Of course, there is no way of knowing for sure that the other person will be honest, but you do have some leverage. If the other person is dishonest, they could be barred from Auction Web. Also, buyers and sellers have the opportunity to comment on their dealings with each other. These comments are visible to anyone who wants to bid on an item, and the e-mail addresses of the people who made the comments are also posted so that you can question them, too. To top it off, E-Bay is quite large—more than one million items have already been sold. Overall, I've seen tons of really neat stuff for sale and I've had some fantastic transactions there (Fig. 1-6). For more information, see:

http://www.ebay.com

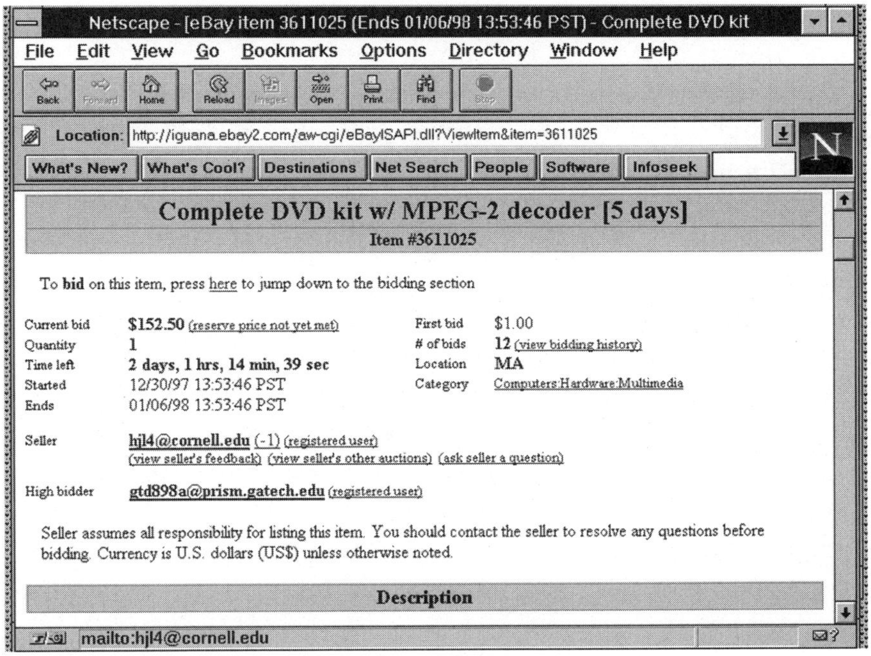

Fig. 1-6 A sample of one of the thousands of items that are up for sale every day on E-Bay on the Internet.

BUYING FROM MAIL-ORDER DEALERS OR LOCAL AUDIO SHOPS

Purchasing used equipment from mail-order dealers was lightly covered in the last section. Most of the mail-order dealers that sell used equipment also have one or more retail outlets. So, when you buy used equipment—even from the same shop—it could be via mail order or from the store itself. Most stores don't have different policies for mail order or for people who purchase locally, but you need to shop in a completely different manner. Most video dealers don't sell used equipment, so this category applies mainly to audio equipment.

If you have the ability to go into a shop and both see and listen to a piece of equipment, you can make a solid decision that is entirely based on your senses, not by putting your faith in what a sales person described to you. By visiting a store, you can thoroughly look over equipment and make some subtle decisions to save yourself money or pick up some more expensive equipment for the same amount of money. For example, if your equipment isn't prominently displayed in your house, you might choose an amplifier that has a few scratches in the cabinet's finish. Audiophiles tend to be incredibly particular about most everything, so if a unit is older and not cosmetically perfect, the sale price should drop significantly.

As mentioned in the previous section, the warranties and money-back guarantees are a major reason why many people purchase used audio equipment from stores, instead of from individuals. After all, it would be tough to sell any equipment at a high enough price to make a profit without some consumer insurance. As a result, the average installation shop or mail-order company that sells used equipment has anywhere from a one-week to a 30-day money-back guarantee. With this standard guarantee, you are usually able to return the stereo component for any reason and get all of your money back, less shipping costs (if applicable).

The average warranty on used equipment lasts anywhere from 30 to 90 days. Generally, if the stereo component fails within the time of the warranty period, you can get your money back or have the unit repaired free of charge. These warranties are usually voided if you have opened the case or cabinet of the unit, physically damaged it, or if the unit was damaged because you connected it to a voltage or to an audio signal incorrectly. For any warranty or money-back guarantee, make sure that you have a hard copy of the signed agreement and be sure to read the specific terms. Some companies will void warranty agreements for reasons that might seem trivial to you. If you void a warranty and can't get your money back, it's your fault.

On the whole, buying used equipment from stereo shops and mail-order companies is a good move. Some shops don't like to bother with hiring a service person to fix trade-in equipment, so you might have a few problems finding what you want, unless you live in a major city. Check your list of shops from the magazines and also check the yellow pages for stereo shops in your area and be sure to call first find out if they sell used equipment. You shouldn't have any problems with buying used and repaired equipment from these suppliers.

PAWNSHOPS AND SECOND-HAND STORES

The flip side of buying equipment from shops or mail-order companies is scouring the pawn shops and second-hand stores. With pawn shops and second-hand stores, you trade a few extra dollars for the security of having a guaranteed unit. You might not need to shell out much money, but you are completely at risk. Another good aspect is that most pawnshops carry plenty of video and home theater equipment, in addition to stereo components.

The second-hand "thrift" stores are generally lousy places to look for A/V equipment. You might find something in one out of every five stores that you look through, but it will usually be an 11-inch black-and-white portable TV from 1983 or something equally undesirable. Few people donate decent A/V equipment to the Salvation Army.

Pawn shops, on the other hand, are great stores to look through for used home theater equipment or other interesting electronics devices. It seems as though when financial hardships strike, the first things to be liquidated are the items of high technology. Likewise, it seems as though the people who need to quickly liquidate their possessions are those who dabble in high technology. If you are taxing your budget to buy your upcoming stereo system, beware of this warning sign! Don't overburden your budget or your beloved system could find a temporary home in a pawnshop!

The two big warnings about buying from a pawn shop are that you won't get a warranty or a money-back guarantee on your equipment and you could easily run across stolen equipment. Be sure to check the physical condition of the piece that you are interested in: Check all of the buttons, switches, and knobs to be sure that they all function. Also look for worn jacks, scratches, dents, oil or grease, burns, and melted areas. If one (or more) of these conditions exist, the price is less than 10% of the original sale price, and you are good with repairing audio equipment, it might be worth your time and money to buy the component. Otherwise, avoid it! You will only lose money on it.

As stated previously, avoid stolen equipment. You can't be sure that anything is or isn't stolen, but if the component's serial number has been scratched or etched out, it has almost certainly been stolen. It is also best to avoid a piece of equipment if the serial number sticker or tag has been removed, painted over, or covered. If you find any equipment that has been altered in this way, complain to the store management and even complain to the proper state or local government bureaus. I have heard that some states have been working to organize and clean up the pawnshops within their boundaries and thus remove the stigma of pawnshops as being a clearinghouse for stolen goods. In these areas, you should have fewer problems with purchasing equipment.

Unfortunately, my location in Pennsylvania seems to be almost devoid of pawn shops. The closest ones to my present location are about 45 minutes away, in different states. But if you live in a city, you should have no problem finding several. A friend of mine regularly scopes out the bargains in three different pawnshops in his locale. If this is your situation, good luck!

HAMFESTS, COMPUTERFESTS, FLEA MARKETS, AND CONSUMER ELECTRONICS SHOWS

The flea markets section at hamfests, computerfests, and consumer electronics shows are the best place to go for absolute minimum prices, with the trade-off of absolute maximum risk. Unfortunately, most people know nothing about these shows and they are missing out on some great buys and interesting technology (Fig. 1-7).

About half of the time when I mention to someone that I am going to a hamfest, the response is a joke, such as "So, are you gonna eat a lot of

Fig. 1-7 A flyer for the spring Timonium, Maryland (near Baltimore) hamfest. With thousands of sellers and more than 17,000 in attendence, this is one of the best hamfests for seekers of audio/video gear.

pork? Ha ha." Yeah, real funny. A hamfest is an amateur radio gathering, where some large, open area (such as a county fairground) is filled with commercial amateur radio venders and another section (usually a section of the parking lot) contains a flea market. The flea market is full of old shortwave and amateur radio parts that range anywhere from brand new to dating back to the turn of the century.

Just a few decades ago, nearly anyone with an interest or expertise in electronics was also an amateur radio operator. Of course, that was before the spread of personal computers and other electronic gadgetry. In many cases, the computer hobby overlaps the amateur radio hobby.

Thus, most hamfests are also computerfests. Because of the overlap between computers and amateur and shortwave radio, you can also find other consumer electronics at these events. Some of the larger events are even advertised as "consumer electronics shows" and all sorts of electronic equipment is present, although the core is still amateur and shortwave radio and computers. The smaller hamfests usually have little more than just shortwave and amateur radio equipment. However, the larger events are really interesting—you have no idea what you might find!

Most of the fests that I have been at recently have been either radio or computer dominated. However, a few large fests that I went to a few months ago had a fair amount of stereo equipment in the flea market. One table featured a number of old tube McIntosh amplifiers. Another contained a big stack of 1980s-era cassette decks in various states of disrepair. One excellent hamfest that I went to last autumn included one vender with a few tables packed full of semi-professional video production equipment. I didn't get a close look, but it included a number of video cameras, some video switchers, digital effects boxes, etc.

The main disadvantage to these fests and shows are that people pull junk out from everywhere and sell it off. After you buy a unit and take it home, it's yours. There is no store or home address to take it back to—just an empty fairground or convention center, thus, the low prices.

But the equipment and the sellers vary widely. I have had everything from very honest and friendly sellers, who were not afraid to tell me anything that was wrong with a unit, to those who flat-out lied about their equipment. Once I bought a cassette deck for $3. When the guy said that it worked fine, I thought "Yeah, right," but the price was excellent and the condition was good. I brought it home and it worked fine! In another case, I bought a VCR that supposedly "worked fine." Instead, the idler wheel was malfunctioning and the VCR would not disengage the tape.

Strangely enough, out of all of the equipment that I have inspected during several years of hunting at hamfests, I have not seen any visible signs of stolen equipment. It seems as though the flea market section, at these type of events would be a great location to sell hot equipment, but apparently they aren't. If anything, they are a great location to unload basement ballast.

For an up-to-date list of hamfests, see the ARRL hamfest event calendar at:

http://www.arrl.org/hamfest

Standard flea markets are not as good for finding stereo equipment. The prices and dealer honesty are about as variable as what you would find at a traditional hamfest. However, the dealers at hamfests are all interested in electronics, so there is a much better chance that you will find mid-grade (or even high-end) consumer audio equipment. At the traditional flea market spot, you will only find low-grade or very low-grade equipment: a $30 (list price) cassette deck in fair condition for $25, a beat-up portable TV to match the one that you saw in the Salvation Army last week, and a pullout AM radio from a '72 Pontiac.

EVALUATING YOUR NEEDS

In order to find the equipment that you want, you must first evaluate your needs and requirements so that you don't wind up with a poor system, a system that isn't useful to you, or a system that is overpriced.

Different people have different needs and preferences in home theater systems. Personally, I'm not fond of the big-screen projection TVs. The resolution is just too poor to appreciate at any reasonable distance. This is beginning to change now that more televisions are being manufactured with display capabilities above NTSC standards. In time, I might like the idea of having a huge projection TV, but right now I wouldn't want anything with a screen larger than about 27 inches.

Aside from knowing what type of sound you are interested in, you also need to determine how long you plan to be using the system. Do you only occasionally listen to music, and fidelity isn't a concern? Then a minicomponent system might be all that you need. However, if you have an appetite for finer sound or you constantly listen to music, invest in better equipment. Not only will it sound more pleasing, but chances are that it will also last longer. If you are an audio head, but don't have the money to buy an entire system at once, buy a system component by component. If you're in high school or college and can't afford the system, listen to a personal or portable stereo until you can save up some money for good equipment. If your equipment lasts a long time, it will save you money over the long run. You probably don't want to sink too much money into a system anyway until you land that all-important first "real" job and are furnishing your own digs.

When purchasing equipment, always check the specifications sheets for information. Compare the fidelity ratings and the warranties. Excellent ratings for fidelity (such as signal-to-noise ratio, S/N, and total harmonic distortion, THD) and long warranties will almost certainly guarantee that the equipment is built well and will survive years of regular use.

One method of choosing a piece of equipment that is often effective, yet doesn't necessarily reflect quality, is merely to pick it up and feel it. In many cases, a heavier piece of equipment is built sturdier (both physically and electrically) than lighter equipment. Chances are that the six-pound VCR is sonically and physically inferior to the 10-pound model (assuming that both were made in the same year). This is because the best equipment uses metal cabinets and front panels for shielding, uses solid and cast mechanical sections for better strength and stability, and uses heavier-grade electrical components.

As mentioned in the previous paragraph, metal cabinets are important for electrical shielding and strength. Except for video equipment, where the emphasis is on price and features, always try to avoid plastic. It is typically a sign of inferior audio quality and that the manufacturer skimped on the physical construction throughout the piece of equipment. Of course, not every piece of equipment in a metal cabinet will last as long as another encased in plastic, but it's a good general rule to follow.

TELEVISION

Television, the appliance of today, is part of an electronics revolution that will place it in the frontier of tomorrow. The basic technical theories of TV and its role as the cornerstone of the home theater system are covered in Chapter 2. HDTV, the system that is destined to replace our standard NTSC system, is explored in Chapter 9. Other associated technologies, such as DBS and satellite TV, are covered in Chapter 10.

VCRS

With DVD technology quickly gaining popularity and the recordable DVD on the horizon, it appears that there will be little room left for the VHS VCR in the future. But for the time being, the VHS VCR is the king of all recorded video formats. Videotape and videotape players and recorders are covered in Chapter 3.

DIGITAL DISCS

The record-sized laser videodiscs from the 1980s were the widespread butt of many technology jokes, including one in the cult movie "Amazon Women on the Moon": "Videodiscs? They're no good. You can't record on them." Today, two forms are vying for market control: laser discs (LDs) and digital video discs (DVDs). Will one of these mediums achieve market dominance or will both follow the same trail of extinction as their ancestor? Look to Chapter 4 for more information.

AUDIO SYSTEMS

Video is only half of any home theater system. Audio is the rest. Audio can be derived from all sorts of mediums, but this book only covers those that accompany video: TV broadcasting, VCRs, and digital discs. Chapter 5 covers improving the audio portion of any home theater system. For more information on stereo systems, see *Home Audio: Choosing, Maintaining, and Repairing Your Audio System* by Andrew Yoder.

WIRES, CABLES, AND CONNECTORS

For years, the key to great sound was to buy great equipment and speakers, and listen to hi-fi recordings. There was no secret to great video. It was simply impossible, unless you bought a projector and films from a theater. By the 1980s, listeners began to question the connections and the cabling that delivered the sound from one component to another. Are the cables susceptible to interference from other components? Are poor connections inhibiting the sound that's reaching your speakers? To solve the problem, gold-plated terminals and connectors, and silver wiring were soon available for those who felt that good wires, cables, and connectors were worth the extra hundreds or thousands of dollars for their system. For more information on the controversy, see Chapter 6.

COMPUTER VIDEO

We're no longer living in 1980. We're at the turn of the century and computers are now *the* topic in the American home. Lately, the hot technological topics with the mass media have been the Internet and multimedia. With the advancements of multimedia home computers since the mid-1990s, the home computer can be used as the keystone of an A/V system. Chapter 8 contains more information concerning computers and video.

ANTENNAS

As we reach the 21st century, terrestrial broadcast television is in a decline, losing tens of millions of viewers to cable and satellite (including DBS) TV. But, just because the technology is declining doesn't necessarily mean that it will be gone anytime soon. Far from it. It appears that standard broadcast TV will be with us far into the future. So, information on antennas to improve reception is included in Chapter 11. Chapter 10 is a light technological look at DBS and satellite television.

SIMPLIFIED EQUIPMENT SELECTION LISTING

The following list was constructed so that you could review it quickly and decide what system would be appropriate for your needs and your home. It is not intended to be a complete guide to comparing all of the many features that are available in audio/video components. If each different feature, such as TV with closed captioning vs. TV with closed captioning and picture in a picture, etc., was covered, the list would be so long that it could not be of value to anyone but the AV expert, who already knows what he or she wants out of a home theater system.

THE BARE MINIMUM

* TV

Options: See other systems.

BASIC HOME THEATER

* A/V amplifier
* Large-screen stereo TV
* Stereo VCR
* Front speakers
* Rear fill speakers
* Subwoofer
* Surge protector

Options: Cassette deck, CD player, turntable, computer with audio/video editing capabilities, camcorder, and DVD player.

COMPUTER-INTEGRATED

* Pentium computer with stereo sound card
* Amplifier
* Stereo speakers
* Video-editing software
* Video-capture board
* Computer DVD player
* Surge protector

Options: Cassette deck, audio/video editing software, CD-ROM recorder, turntable, and camcorder.

THE VIDEOPHILE DELUXE

* TV receiver
* Video projector
* A/V receiver

* DVD player
* Speakers
* Specialty equipment racks and stands
* Specialty cables, cords, and connectors
* A power-line conditioner
* An acoustically "perfect" room with plenty of seating
* Plenty of movies
* A theater-style popcorn machine

Options: CD transport, D/A converter, turntable, CD recorder, dedicated amplifier or two mono amplifiers operated in stereo, and preamplifier or passive-control unit.

CONCLUSION

Thousands of different stereo components are available, and without keeping abreast of the recent information, you can easily get lost in the sheer numbers. Many fewer manufacturers are presently in the TV business, but this is bound to change as computer companies become involved and the TV specifications change.

Remember, the keys to success in purchasing most anything are good information, a level head, and discernment. Don't be blinded by either amazing, expensive brand-new technologies or by dirt-cheap prices. Buy the best quality equipment that you can for the money that you are willing to spend.

2 Television

Like it or not, TV is at the center of modern American culture. Whether you grew up watching "Howdy Doody" and "Captain Video," "Bullwinkle" and "Banana Splits," "Scooby-Doo" and "Grape Ape," "Smurfs" and "Strawberry Shortcake," or "Pinky and the Brain" and "Ren and Stimpy," chances are that parts of your life—since you were very young—have been dramatically affected by the tube.

I can personally attest to the power of television. Some aspects of my life are certainly vastly different than they would have been without it. For example, some of the inspiration behind my career choices was based on some things that I saw on TV. That's not to say that the masses of the world are necessarily led around by TV producers. I've probably been inspired just as much (or more) by TV segments, programs, or ads that I disliked or disagreed with.

Television impacts our lives so powerfully because it combines our two most powerful senses (sight and sound) into one medium. Reactions to television are typically strong, immediate, and very emotional. Just like life, television's sound and images zip past. By contrast, books and other print media produce a slower reaction, one that is more permanent, but able to affect a much lower percentage of people. Radio occurs in real time, like TV, but most people seem to believe that audio is much more powerful with images--hence the cancellation of all major radio comedies, dramas, etc.

Light television history

The concepts of TV go back to the beginning of the century, as radio was being developed as a viable communications medium. It just made sense that if audio could be transmitted through the air, someday, both audio and video could be transmitted. Many people, not just inventors, fairly accurately predicted television, the equipment, and the potential programming (Fig. 2-1). The problem was technical: How do you convert a moving image into an electrical signal, then convert it back to an image on a screen of some type?

Fig. 2-1 An amazingly accurate "look into the future" of television broadcasting from a 1920 issue of *QST*, an amateur radio magazine. QST

Several of the top radio companies struggled to devise the equipment that would result in millions (or billions) of dollars in future profits. These companies were still in an all-out patent war for the rights to radio and they knew that a complete victory (not just having a few minor ideas of the TV system patented) was essential to reap the fortunes of their inventions.

These companies developed some rudimentary television systems that were publicly displayed at times through the mid to late 1920s. John Baird created a mechanical television system in England that was so successful that the BBC began regular TV broadcasts in 1929. Unfortunately, the Baird system produced a noticeable flicker and the picture resolution was much poorer than today's NTSC standard. Baird's system nearly paralleled that of Charles Jenkins in the United States, except that Jenkins had no great broadcasting system for support—just his own little, experimental TV station. Likewise, AT&T and General Electric also experimented with television broadcasting, and by 1929, a number of experimental, licensed TV stations were operating in the United States—with virtually no viewers.

While some of the most powerful companies in the world spent millions of dollars employing the brightest engineers with the best facilities, something strange happened. A poor farm boy from Utah developed the key component of the television system. At the age of 19, Philo Farnsworth was forced to drop out of the Engineering program at BYU because, with his father's death, the family was no longer able to afford the tuition. A year later, he started a radio shop in Salt Lake City that quickly failed. By the next year, he had developed the "image dissector," the key component of electronic television cameras.

RCA took the greatest interest in television and David Sarnof, president of the company, estimated that RCA spent $50 million dollars in electronic television development in the 1920s and 1930s—a tremendous sum, when translated into 1998 dollars! RCA could almost complete a high-quality television system, but not without Farnsworth's image dissector, yet RCA refused to license its use from the inventor. Instead, they proceeded. In the 1930s, a court ruled that RCA used a component so similar in design to the image dissector that they would not be permitted to use it. Farnsworth refused to sell his patent, so RCA had no choice but to pay royalties for each TV manufactured using this equipment. One widely reported story concerning this event (in both radio/TV and business circles) is that RCA's patent lawyer was so distraught about paying licensing fees to another company that his eyes welled up with tears when it was his turn to sign Farnsworth's agreement.

Although the drive for technological superiority and incredible profits certainly speeded the development of television, it also prevented wide-scale broadcasting for a number of years. If RCA and Farnsworth could have collaborated on a system in 1929 instead of 1939, perhaps millions more people could have enjoyed the medium. For example, imagine if the system had been assembled, TV stations would have been broadcasting, and TVs would have been ready to sell by 1932. Although they would have been expensive, possibly millions of TVs could have been sold by the end of the decade (as it was, thousands of the mechanical TVs were sold in the United States in the 1920s).

If TV could have developed this sort of stronghold in the United States in the 1930s, perhaps it could have been active through World War II. Today, many elderly people recall sitting around their radios and hearing President Roosevelt's speech about the attack on Pearl Harbor. Instead, they could have been watching the *Oklahoma* sink. Similarly, with Hitler's love of propaganda through all forms of media, it would have been interesting to see if Nazi Germany would have adopted the TV technology and sold low-cost "volks TVs" across the country to further propagate the propaganda. The possibilities are fascinating.

Instead, the modern age of TV began in 1939/1940, when RCA and Farnsworth came to terms and new TV stations and receivers were sold (Fig. 2-2). Of course, World War II froze the production and sale of consumer goods, so TV was on hold until after the war. Finally, by 1947, life in America was getting back to normal and TV was back on course to stir the public's imagination. TV sales and the numbers of TV stations skyrocketed during these early years. But, just like the Internet and DVD, it took a few years before TV was everywhere.

In the late 1940s and early 1950s, TV was somewhere between an experiment and a well-matured medium. True, the technology was new and the broadcasters weren't quite sure what would work. However, the networks, stations, and personalities had 25 years (or more) experience in radio broadcasting. It made sense that what was successful on the radio would also be successful on television. So, many of the radio shows, such as "You Bet Your Life," "Amos' n' Andy," Jack Benny, "The Lone Ranger," "Meet the Press," "The Steve Allen Show," and "People are Funny" crossed over to television with varying degrees of success.

Fig. 2-2 One of the early television receivers, a 1938 model from DuMont, which later created a short-lived broadcasting network.

The biggest difference between television in 1950 and that of today is that it used to all be live. When you consider the difficulties of live television, it's amazing that everything was performed that way. Most actors and actresses had prior live radio or vaudeville experience, so the shows were surprisingly well performed. Still, there were occasional mistakes and ad libs, which must have made TV viewing even more exciting. The reason for live television was that videotape was still in the developmental stages in the early 1950s and film was both expensive and it required more time for development. And that's one real advantage to TV viewing today: We have 50 years of the best TV to watch as reruns. In 1950, there were no syndicated reruns to watch!

Television changed at a much slower rate after the development and implementation of videotape recorders in the mid to late 1950s. The 1960s really defined what television was and what it would become. Animated shows, such as "Rocky and Bullwinkle" (created by Jay Ward, who also created the first TV cartoon, "Crusader Rabbit"), "Yogi Bear," the "Flintstones," and the "Jetsons," were tremendously successful. Other developments in general programming included more intricate set designs (not just one room), outdoor shooting (such as "Dragnet," "Gunsmoke," and "Adam 12"), and more complexity ("Twilight Zone" and "Outer Limits"). But perhaps the most significant change was the manner in which television news was reported. Instead of simply being radio news with a few video clips tossed in, television reporters actively sought out their stories. TV cameras were everywhere from the front lines of Vietnam to space vehicles on the moon.

The 1970s and very early 1980s were primarily just a continuation of what had happened earlier. The biggest change in television actually began to occur in the early 1970s, but was not widely felt until the 1980s. In 1972, a small company from Wilkes-Barre, Pennsylvania, called *Home Box Office*, began offering programs, such as NHL hockey games, to a few hundred cable-TV viewers in the area. These programs were all transmitted via microwave, so it was a strictly regional service, with little hope of ever catching on nationally. But all of this changed in 1975, when HBO broadcast its first program via satellite. Within two years, Ted Turner was getting his poor channel 17 TV station on the satellite. In several more years, WTBS, once the lowest-rated station in the Atlanta market, was watched by millions.

Throughout the end of the 1970s, satellite-broadcast programs grew steadily, but failed to attract a high percentage of the TV market. This was because cable-TV systems were still only serving a small percentage of viewers and the available cable stations were still in the process of carrying the new satellite-fed programs and networks. Because of the slow growth, most future satellite TV entrepreneurs were slow to "jump" into cable TV. Everything changed early in the 1980s, when CNN, MTV, ESPN, and many others jumped in. By the end of the decade, nearly all cable TV systems carried satellite-fed programming, and cable had grown to serve approximately half of the people in the United States. For more information on satellite TV, see Chapter 10.

Aside from new technology and that people had to pay for their programming, the big change here was the diversity and sheer volume of programming offered. For example, music videos and live concerts had received limited air time in the past, often confined to spots on little UHF stations after midnight. With MTV, you could watch pop and rock music videos all day. Likewise, ESPN brought a mass of sports to the table; not only could you watch some sports that received lesser coverage (such as hockey and basketball), but you could watch this programming at times when no other networks were covering it.

The 1990s have been somewhat like the 1970s in terms of the changes in television broadcasting. No dramatic changes have occurred, aside from the fairly limited development of direct-broadcast satellite programming. Instead, new technologies are being developed, some of which will certainly be implemented in the first decade of the next century. Will DBS satellites take over entirely from terrestrial broadcasting and cable TV? Will HDTV be successfully implemented? See Chapter 9 for more information on HDTV. Will TV be broadcast at acceptable quality via the Internet? Or will some new interactive TV system that merges television and computer technology find acceptance?

TV TECHNICAL BACKGROUND

Television is a very complex technology because it combines the mediums of audio and video transmission. In fact, it is often easiest to divide television into its audio and video components and discuss each subject separately.

The purpose of this section is not to provide a detailed look at the technological aspects of television—don't expect to suddenly be able to draw up television schematics or design a receiver. Like the chapters on digital video and videotape, this technical background is intended to provide information so that you can make better decisions for purchasing equipment and, when necessary, having it repaired. Especially, these technical specifications and theories are important for understanding the approaching high-definition TV system. For more information on television theory, see *Electronic Communications (6th Edition)* by Robert L. Shrader.

AUDIO

The audio component of television is simplest to describe. Theoretically, it is nearly identical to that of an FM radio station. The FM radio or TV station transmits a basic signal. This signal remains on one frequency until sound runs into the transmitter. Then, the frequency of the signal jumps above and below the formerly steady frequency. This distance between the signal's maximum frequency and the minimum frequency is known as the *deviation*. The deviation is the only difference between FM and TV broadcast audio. FM deviation is 75 kHz and TV audio is 25 kHz (Fig. 2-3). If you have a local channel 6, you can hear the difference on your FM radio: Even if channel 6 is very strong, the audio, as heard on your FM receiver, will still sound weak by comparison to the other FM stations.

Fig. 2-3 A comparison between the amplitude of a TV audio signal (left) and an FM broadcast signal (right).

The deviation of an FM or TV broadcast signal is equal to the loudness of the audio. The greater the deviation, the louder the audio.

TV audio broadcasting is very complicated—especially when you get into stereo transmissions, stereo indicator signals, and SCA. Because of this and because explaining the rest of the audio complications will do little to help you choose or use a TV, the technical aspects of TV audio are not covered further in this book.

VIDEO

The key component of the television signal is the video portion. Without the audio portion of a television broadcast, the signal would be silent TV, but TV nonetheless. Without the video element, a television broadcast would be a radio transmission (Fig. 2-4).

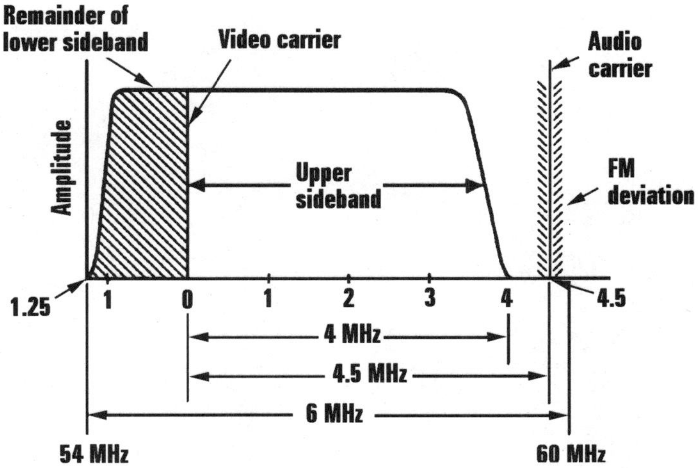

Fig. 2-4 A view of the TV broadcast signal, showing the width of its various portions.

SCREENS, SCANNING, AND DPI

When you take a photograph, the light travels through the lens and leaves a permanent registration on the film. When the shutter opens, the light leaves its mark on the entire area of the photo (image area) at once. Because of the nature of video monitors and broadcasting, it would be nearly impossible to implement a system whereby the video would appear simultaneously on the screen. Instead, video is transmitted dot by dot, line by line, just like photographs printed in a book, magazine, or newspaper. Because fewer dots per inch are used in newspaper photographs (typically at about 100 dots per inch), the dots are much clearer to see than in books or magazines (which are often as many as 2400 dots per inch).

The more *dots per inch (DPI)* used in a picture, the clearer, sharper, and more detailed it will be. A large DPI is always best for the final result, but in many instances, it isn't feasible. In the case of newspapers, the low quality of the newsprint prevents clear printing at a high DPI. In the case of Web pages on the Internet, high-DPI pictures require much more information, hence much longer time for

Fig. 2-5 An example of the difference in resolution: At left, a photo at 280 DPI. At right, the same photo at 75 DPI.

them to download and appear on the screen (Fig. 2-5).

The transmitted video signal consists of an area of 525 dots wide by 495 lines high. By comparison to photographs, printed artwork, and theater movies, television has a very low resolution. As mentioned concerning newspapers and Web pages, there were also some powerful tradeoffs between high- and low-resolution television during its developmental stages.

SETTING THE STANDARDS FOR TV

When the *National Television Standards Committee (NTSC)* began developing the standards for video broadcasting in the United States in 1940, they realized that they had to create a system that could be effectively implemented. A great system that was too costly (either to broadcasters or consumers) would fail and doom the medium. The problem with transmitting high-resolution video wasn't so much that it couldn't be done; it was the required signal width. If the video signals were too complex, each channel would require so much space that only a few channels could be used (instead of the present-day 69 channels).

To solve the space problem, the NTSC tested television on many different viewers. They showed the video at different resolutions and chose the lowest resolution that the viewers would accept. From this test, they determined the resolution for television. Based on these determinations, the *Federal Communications Commission (FCC)* established how the channels would be aligned and how much space each would require.

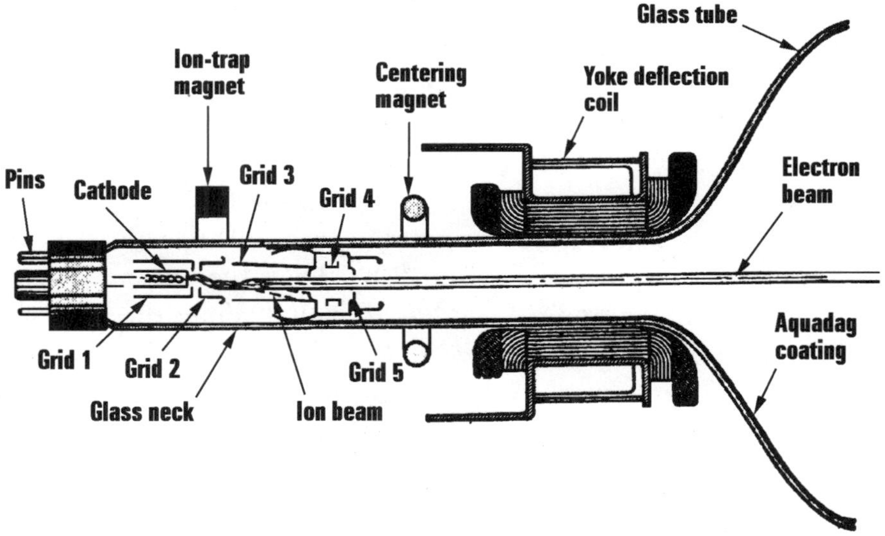

Fig. 2-6 The components in the CRT neck. Follow the path from the grids and the cathode out to the end of the tube.

The picture appears on the screen dot by dot, but as noted, it is not feasible for everything to appear on the screen at one time. Instead, three different electron guns are mounted in the back of the picture tube (*cathode ray tube*). Each of the guns shoots a ray of electrons to the front of the tube. The inside of the screen is coated with a phosphorescent material that will glow when it is energized. Each gun shoots a different-colored beam (either red, green, or blue) to the front of the tube, where it energizes a phosphorescent dot on a layer inside the front of the screen. (Fig. 2-6).

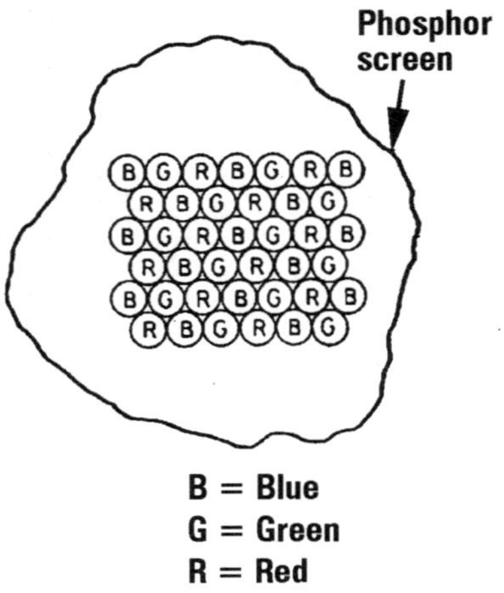

Fig. 2-7 A simplified view of the multicolored dots on the phosphorescent coating on the inside of the picture tube.

The CRT coating is applied in a pattern of dots (Fig. 2-7), with groups of red, green, and blue dots that correspond to the red, green, and blue electron guns in the neck of the tube. The guns fire on each set of three color dots, scanning from left to right, as you would read the lines in a book (Fig. 2-8). Then the gun moves down to the next line and scans it from left to right. When the screen is finished, the process starts all over again. The dot patterns are fired so fast that you can't even see the changes in the picture, let alone in the line-to-line scans.

With the electron guns firing patterns faster than you can see, it's probably no surprise that the guns don't have to change position for each different set of dots that are fired upon. Instead, the direction of the beams are changed magnetically with a coil of wire wrapped around a circular set of magnets that are placed around the outside of the glass neck of the CRT. This assembly is known as the *yoke coil* or *yoke deflection coil*. As the scanning beam passes from line to line, the magnetism fluctuates, bending it so that it can continue on its journey.

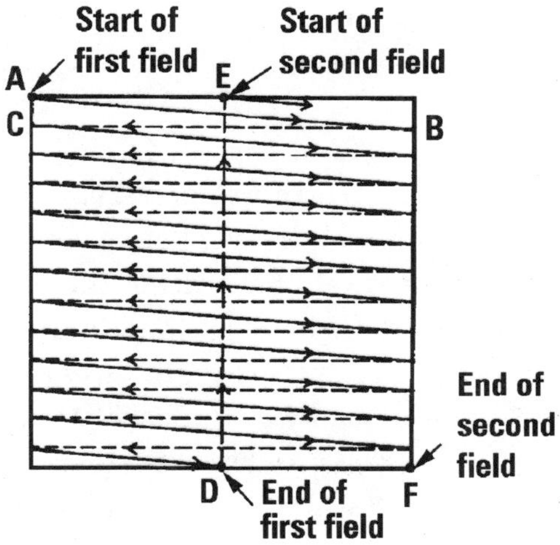

Fig. 2-8 An example of the lines scanning from side to side in an interlaced video image.

In addition to bending the beams as they travel through the CRT, the yoke coil also helps establish the boundaries of the television picture. If the circuits that control the yoke coil are working incorrectly, the picture might extend off the screen or the screen might be divided. For example, when I was very young, my Grandma had a television where the top several inches of the screen's images were stretched out and a bit bent to one side. Everyone's head looked about a foot longer than Abe Lincoln's. Chances are that the TV had a problem in the vertical deflection circuits, which produced an incorrect picture size and shape.

FRAME FREQUENCY

Another important specification for the way that moving pictures are presented via television is the *frame frequency*. This term is the number of times that a new picture is scanned on the screen. To present video with unnoticeable flicker, the frame frequency must be high. In the NTSC television system, a new frame appears 30 times per second. The picture quality could be improved if the frame-frequency standard would be increased, or reduced if the standard would be decreased. For example, amateur radio operators use a narrow-bandwidth television system on the shortwave amateur bands: *slow-scan television (SSTV)*.

SSTV is so slow that the video appears as a series of pictures. Instead of a number of frames appearing per second, one frame appears every eight seconds!

The problem with running 30 frames per second is that some flicker is still noticeable. The flicker could have been removed by increasing the frame frequency, but that would have made the TV signals wider (eating up more of the radio spectrum) and increased the amount of film and tape necessary for each program (greatly increasing the cost for station operation—especially in the early days of television). Video engineers worked around the problem with a clever deception. Rather than showing one frame at a time and having a harsh break from picture to picture, they transmitted just the odd-numbered lines (with the even-numbered lines blank) and then just the even-numbered lines (with the odd-numbered lines blank). This system effectively smoothed out the transitions from picture to picture, without increasing the cost of the television production.

This practice of transmitting blank lines to smooth the frames is called *interlaced video*. Interlaced scanning has been used for nearly all types of electronic video since its inception. But in the 1980s, computer companies broke away from the old standards. Up until that point, you could plug your computer into your TV and use it as a computer monitor. Then the computer companies began manufacturing systems with non-interlaced monitors that could only be used with the new computers (which were not compatible with the interlaced NTSC TV system). This was no big deal because at that point, home computers had become sophisticated enough that they weren't just a toy to play with by the TV. However, as is described in the HDTV section, this change could dramatically affect the future of television.

SYNC PULSES AND AGC

One key ingredient to the television signal is neither audio nor video, yet it is essential. *Synchronizing pulses* (known as *sync pulses*) are simple tones that key every time that the frame must change. The television timing circuits use the sync pulses as a signal to start scanning at the upper left-hand corner of the screen again. Without these pulses, the TV would not "know" when to start scanning a new picture, rendering the video unviewable.

Yet another problem stood in the way of the development of a functional television system. How can you set the brightness of a signal? The video signal is constantly varying in brightness (strength), so the receiver can't "look" at the signal strength and automatically set the brightness on that characteristic alone. Otherwise, it would either vary while you were watching the program or it would vary, depending on when you tuned in.

Another possibility would be if it would take a second or two to sample the signal and set the high point in the signal as white and the low point as black. This system is also doomed because stations vary in strength—if this system was used, a strong station would be too bright and a weak station would be too dark.

To properly set the brightness of the television, the engineers needed to find something to "lock" on to—a signal that would be set at a constant level at every TV station. They cleverly chose the sync pulses to calibrate the brightness. A circuit called the *automatic gain control (AGC)* monitors the sync pulses, "notices" if the signal strength raises or lowers, and compensates appropriately. This system is effective because the sync pulses should be broadcast at the same level at every television station, so no matter if the signal is weak or strong, the AGC circuits will notice the difference in sync pulse strength.

AGC circuits are also used in radios to prevent the stations from sounding very quiet during a signal fade, then suddenly becoming very loud as the signal improves. Of course, radio stations need no sync pulses, so the AGC monitors the signal level and varies the output accordingly.

SCAN LINES

As mentioned earlier, television stations transmit a number of dots that form lines on the screen. Typically, about 490 vertical lines are transmitted by 525 dots in width. These numbers are more than just technical gibberish; they determine an important video factor: resolution. These figures also affect the aspect ratio.

Resolution, the amount of video information in a given area, was mentioned earlier. For several decades, the resolution of NTSC video has been fought over as organizations lobby for HDTV. The fight hasn't been whether or not HDTV should be approved—everyone seems to be in agreement that NTSC's resolution is deplorable. Instead, the fight hinges on what company or alliance of companies should be approved for determining the HDTV specifications.

The *aspect ratio* has nothing to do with the quality of the picture that you see. Instead, it is the actual shape of that picture. Because the image area of the CRT has four sides, you can think of that picture as being a rectangle. With a resolution of 490 (vertical) by 525 (horizontal), the shape of the picture is rectangular and the width is a bit wider than the height. The standard NTSC aspect ratio is about 4:3. A few problems result from this aspect ratio:

* The human range of vision is much wider than 4:3. At 4:3, the impression is very box-like.
* Movies that are produced for the box office also have a much wider aspect ratio. To fit them onto a television screen requires a large amount of manipulation, often to the detriment of the motion picture.

The present-day solution to the aspect-ratio problem isn't much of a help, either. Some television manufacturers and movie-production companies produce equipment and films in the *letter-box format*. In this system, the aspect ratio is kept the same as in the theater by filling the width of the television screen with the width of the letter-box video picture. This system is fantastic if your television is equipped with a 16:9 format screen.

Unfortunately, the transition from a letter-box video to a 4:3 ratio screen isn't so smooth. Because the aspect ratios are so different, the height of the letter-box picture is much smaller than the height of the screen. As a result, the rest of the screen is unused and filled with black—significantly reducing the viewable image area of your television screen.

One variation on the theme of aspect ratio involves a few novel little tweaks. For example, some Pioneer televisions have different settings to purposefully distort the video image (Fig. 2-9). The concept here is that a slight picture distortion, if properly performed, could heighten the effect of whatever you're viewing. I think that it would be great if they added a few other screen variations, such as "fun house mirror" effect, to increase my enjoyment of "Baywatch."

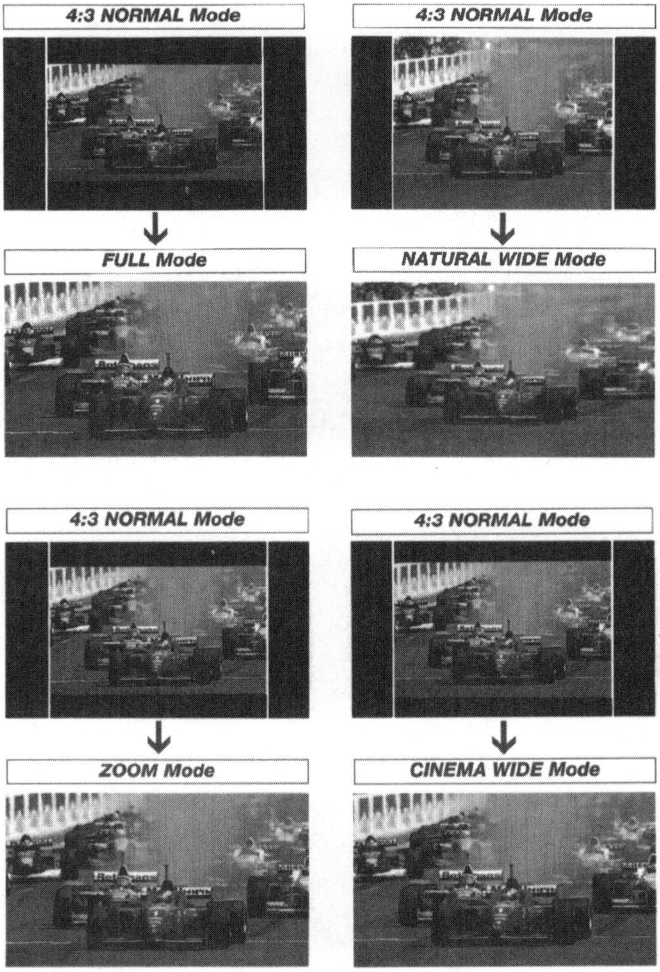

Fig. 2-9 The four Pioneer 16:9 screen variations. Pioneer

TV CHANGES

As has been alluded to, the NTSC format is technologically outdated. The resolution is too blurry, the audio is subpar, and the aspect ratio is misshapen for viewing movies. This isn't just my own personal bias; viewers are unanimously in favor of positive changes to these qualities.

The dilemma is not whether to make the changes, but how. Some of the NTSC standards could be altered to improve the quality. Perhaps the new TVs could be made so that you could simply switch between NTSC mode and higher-resolution video. Or maybe the aspect ratio could be changed and older TVs would receive the signal in a letter-box format, but new TVs would be manufactured in the wider aspect ratio.

Instead, none of these options are being pursued. The NTSC standard that has served for nearly 60 years is being tossed out for an entirely new system that is improved in all of the aforementioned ways and is transmitted digitally. The electronics manufacturers have been struggling to get a system in place for more than a decade, but for broadcasting, the system must be approved by the agencies in charge of communications (the Federal Communications Commission in the United States, the Department of Communications in Canada, etc.). The FCC has taken a "let the marketplace decide which technology is best" approach and this strategy appears to have significantly slowed down the approval of a new broadcasting system.

Finally, after years of deliberations and different strategic alliances, the new system (HDTV) was announced in 1997 (for more information on the specifics, see Chapter 3). Also, in 1997, several of the largest computer manufacturers banded together and put forth a television proposal that was entirely different from the HDTV system that had taken so long to finalize. This proposal differs in the technological specifics of the picture and sound, and also in the actual content of the programming. Rather than just television, these companies are pushing for a system that would be integrated with computers and the Internet.

Instead of just watching a program, you would be able to comment on it and even order products directly through your TV.

The problem is that although the proposed Internet/television combination is very tempting, how long will it be before the standards for this integration can be agreed upon? As it is, the relatively simple upgrades to the HDTV system required years to be ironed out. Can we expect any less time during a dramatic switch to an integrated computer/television system? For more information on computer video, see Chapter 8.

Even though the standards for a proposed broadcast standard have been passed, they won't necessarily take effect. A large-scale case of this bureacratic tie-up is the expanded AM broadcast band in the United States. In 1995, the FCC approved a ruling that AM radio stations could begin broadcasting in the region between 1600 and 1700 kHz. The 10 radio channels were assigned to a number of different broadcasting organizations around the country. But, some of those organizations that applied for expanded-band licensed felt that they were treated unfairly and they filed against the FCC. The FCC, in turn, froze the acceptance of all stations that had not started broadcasting before the complaints piled in. As a result, at this writing in late 1997, of the dozens of expanded AM band stations that were licensed by the FCC, only WJDM on 1660 kHz is regularly operating.

You might be wondering where this section is going. After all, how do these government standards apply to you? The problem is that the equipment now required to receive and play-back television signals could be rendered useless within 10 years. However, this all depends on whether the plan unfolds as the FCC believes it will. If the progress inches along at the pace of the opening of the expanded AM broadcast band, it could be decades before the NTSC television system is finally replaced--whether it is with the accepted version of HDTV, DTV (the computer-integrated system), or something else. That is why a chapter has been dedicated to NTSC television in this book. If the stations had been broadcasting HDTV when proposed, NTSC would already be antiquated.

The bottom line is that some form of improved digital TV is on the horizon. According to the FCC and the HDTV equipment manufacturers, it will be available in late 1998 at the earliest. But if you want or need a new television before the turn of the century, chances are that you will safely have at least five years of use (possibly much longer) before you feel compelled to make the jump to a new video technology.

EQUIPMENT ON THE MARKET

Because we are still thoroughly entrenched in the old system, the television-reception equipment on the market still follows NTSC standards. However, some equipment is now manufactured to accommodate the better specifications of nonbroadcast media, such as DVD and laserdisc players. The following sections cover the latest pre-HDTV equipment.

MODERN TELEVISIONS

After approximately 60 years of experimentation, the NTSC system has undergone only one major change (color). Obviously, the manufacturers have had plenty of time to perfect television receivers. At this point, what can possibly be done (Fig. 2-10) to further refine the good ole TV?

These days, few changes are being made to the performance of the television. You might even find older sets from the 1970s that perform just as well, or possibly better than some of the newer models. Instead, the manufacturers are miniaturizing the circuitry and adding little gadgets to make TV viewing a bit more entertaining, if nothing else.

The miniaturization of circuits is a great help for those looking to purchase a new TV. Although smaller circuits do not equal better performance, they do mean lower prices. Early on, televisions were assembled part by part, with plenty of large, heat-producing vacuum tubes throughout. These TVs required regular maintenance for their vacuum tubes. Also, there were so many parts that literally anything

Fig. 2-10 One of the more-clever methods of improving the small screen size of televisions over the years, the Magnaviewer. If you can't make the screen any larger, just stick a really big magnifying glass in front of it!

could go bad. Today, most TV circuits have been consolidated into very small integrated circuits, with few extra components. As the number of ICs produced increases, the prices drop. And if an IC fails, it can be replaced with another low-cost IC.

FEATURES

As mentioned, aside from improvements in video-screen resolution and S-VHS inputs, recent advances in present television receiver technology have primarily been limited to miniaturization and gadgetry. This section covers some of those handy gadgets that might not make the picture look much better, but they sure are fun to play with (Fig. 2-11)!

Picture in a picture (PIP) With picture in a picture, the receiver can display the video from two different channels at the same time. One picture is displayed that fills the entire screen and the audio from that channel is output to the speakers. The second channel is displayed in a small square that appears over the top part of the

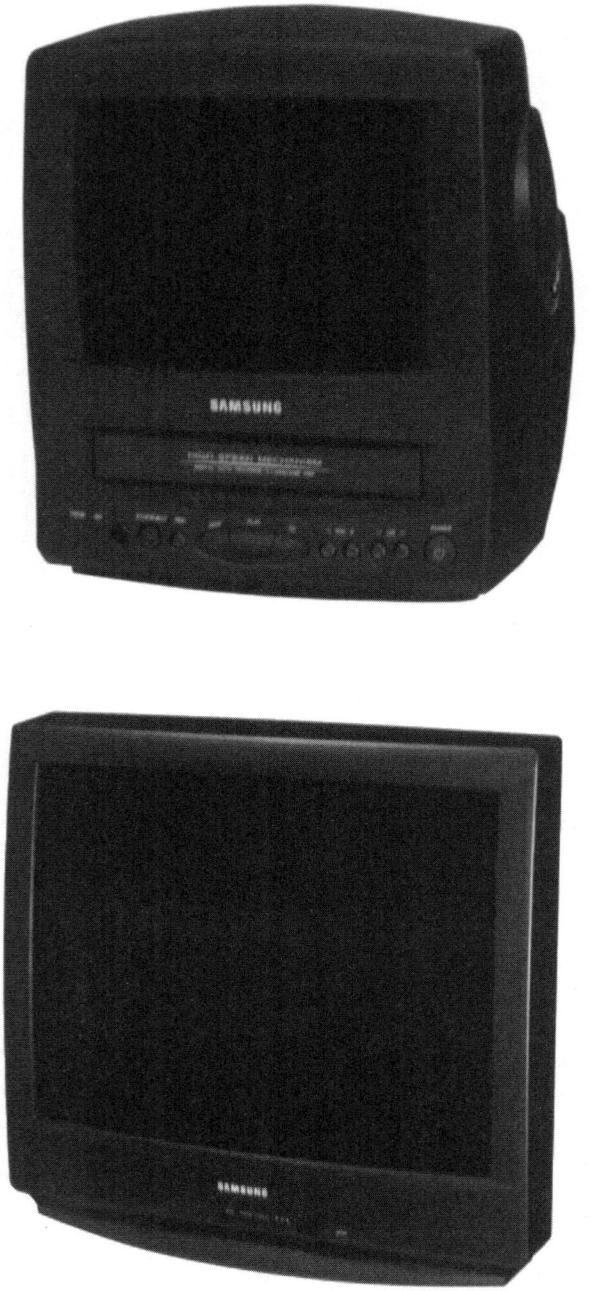

Fig. 2-11 A portable TV with a built-in VCR, perfect for getting "back to nature" while camping out (top), and a Samsung.

Fig. 2-11 (continued) And the screens keep getting bigger. At the top is a 35-inch TV with the ever-popular picture-in-picture and parental channel lock features. At the bottom is a high-end projection TV. Samsung (top) and Runco (bottom)

first picture. In some TVs, the size and location of the smaller screen can be changed.

High resolution Although the resolution of the NTSC video signal can't change, the resolution of the video output can be improved, provided that a higher-resolution signal is inputted to the TV. For example, if you connected a DVD player to a TV with better-than-NTSC resolution, the resolution would also improve. Manufacturers also have been looking at the possibility of producing televisions that can be switched between various resolutions for TV reception flexibility (so that a TV can be used to watch the different types of digital TV that are being created. See Chapter 9 for more information on HDTV.).

Ac/dc operation This option is only available in the less-expensive portable TVs. With the setting on ac, the TV can be operated from 120-Vac power; on dc, it can run on 12- to 13.8-Vdc power, which is available from the electrical systems of most vehicles.

Audio power This specification is the amount of audio power that the television can deliver. However, for most people interested in home theater, this specification is useless. It is better to have the audio amplified by an audio amplifier or a home-theater amplifier than by the TV.

Inputs The inputs to a TV provide a way for the signal (from an antenna, cable-TV system, or video player) to be viewed on a TV. On older TVs, the only inputs were either two 300-ohm screw terminals for antennas or a UHF-type connector for cable TV. Most modern TVs include the UHF-type connector and also separate RCA jacks for left and right audio input and video input.

Outputs As with the inputs for modern TVs, the outputs have also changed significantly. Sets from 20 years ago typically had no outputs. Today, most TVs have separate RCA jacks for left and right audio output and video output.

Closed captioning The closed-caption system provides text on the screen that follows the dialog of the program. Closed-caption TVs are an obvious boon to those who are hearing impaired.

Screen types In this case, "screen types" does not mean the resolution of the screen, but rather the actual type and characteristics of it. For example, modern televisions have very flat screens with sharp corners. Older models were much more rounded, both in the corners and in the plane of the screen. The greatest modern television screen development has been that of the wide projection screen. Large CRTs are extremely expensive to produce. To avoid the cost, manufacturers began using a system where the three color guns (described in the first section of this chapter) each shoot a tiny beam of light (not electrons) against a translucent screen that is stretched across the front of the set. The problem with projection TVs over the past 10 years has been the NTSC resolution. The number of lines available on a picture does not change, no matter what the size is. So, each dot on a small screen will appear much larger on a much larger screen. The end result is that the projection TVs looked bad—blocky and blurry. But because modern, higher-resolution formats (such as DVD) can now be displayed on projection TVs with better-than-NTSC resolution, they can be utilized more effectively. In fact, with the advent of HDTV and the increasing acceptance of DVD, it appears that wide-screen projection TVs will become increasingly popular, possibly even eclipsing the sales of CRT TVs in a few years.

Stereo Now, most new TVs with screens larger than 20 inches can receive stereo audio. The idea here is that any TV with a screen smaller than 20 inches will only be used in portable or secondary viewing applications (such as in a kitchen), where stereo audio is not a real need. But, you can assume that any new TV that is manufactured for primary viewing will be capable of receiving and playing back stereo audio.

Channel lock Channel lock is an interesting technology for parents who receive cable TV. If parents don't want their children to be able to view certain channels, they can enter an access code and render certain channels unviewable. If you have kids and a TV with channel lock, you can enjoy blocking channels and hearing your kids tell you

everything they know about the First Amendment.
Ratings lock All newer TVs with a screen size larger than 13" are produced with the dreaded V-chip—the circuitry that allows parents to enter a code and restrict all programming below a predetermined rating. The TV stations broadcast the ratings along with the programs being transmitted and the V-chip is capable of determining the rating and blocking the signal, if necessary. If you have an older TV that existed before the V-chip, you can buy a separate box that you can set to block certain programs. Time for even more First Amendment fun.

Channels How many channels do you really need to receive? For terrestrial TV, stations can broadcast on 2 through 13 (VHF) and 14 through 69 (UHF). In the United States, the UHF band did extend to channel 83, but channels 70 to 83 were eliminated and the frequencies were sold in an FCC auction. Many of the cable TV systems around "only" have 40 to 60 different channels, but some of the systems in large metropolitan areas offer more channels than you can shake a remote control at. To handle all of the possible channels, most of the new TVs offer 181 channels, which should be sufficient!

Channel scan The channel-scan feature enables you to press a button and the television scans through the channels, pausing for a few seconds at each channel.

Clocks and timers Clocks and timers are handy little gadgets that can be used to control when the TV turns on and off, or simply display the time on the screen. The timers are handy if you regularly fall asleep in front of the TV and hate waking up to old reruns of "Star Trek."

Comb filters Unlike most of the gadgets listed in this section, comb filters are legitimate components of a TV. The comb filter is used to clarify the signal. Because it filters the different frequencies, it improves the horizontal resolution and reduces the amount of bleed-over color. Most of the better televisions include digital comb filters, and this is one of the few features that you should seek out.

S-video input If you decide that standard VHS VCRs just don't supply the video necessary to keep you glued to the set, you might consider an S-VHS VCR, which can provide a better-resolution picture than a standard VCR. Of course, you will also need a TV with S-video inputs.

Speakers Most televisions have at least one built-in speaker so that you can just turn on the TV and enjoy. But some televisions include higher-power stereo speakers in the cabinet. A television cabinet is certainly not an ideal location for speakers and stereo separation. If good audio is a concern, don't worry about stereo speakers built into the set; just buy the TV that you want, and purchase separate speakers and an amplifier later.

MONITORS

Most equipment that you watch video on is either TVs or TV/monitors. But, some are simply monitors. A monitor is a TV that lacks the receiver portion. Thus, it is only good for playing back video from a recorded medium, such as a VCR, laserdisc, or DVD. Monitors are cheaper to manufacture than TVs, so they are attractive if broadcast signals are of no interest. Monitors are rarely used for personal use. Instead, they are often purchased for institutional use, such as in schools, churches, and companies. Another handy application of monitors is in conjunction with a computer, cable-TV and/or TV receiver board (Fig. 2-12). For more information on computer video, see Chapter 8.

PROJECTORS

If you are in the mood for some serious video entertainment, you might consider a trip to the local theater. There, you can enjoy the movie on a huge screen with excellent fidelity. Of course, the movie costs $7 (or more) per ticket for admission, a tall woman with big hair is sitting between you and the screen, the kid behind you is making rude comments throughout, and the guy beside you has bad gas. Was it really worth the trip?

Fig. 2-12 An example of the convergerence between television and computers. Here a television-compatible computer monitor is being used for video conferencing on the Internet. Those involved appear to be quite pleased with the results. Princeton Graphics

These days, you can just as easily have a complete theater system in your home, although it can be rather expensive. In fact, depending on the system that you choose, it could cost more than your house!

The center of a home theater is the video projector. This type of video projector is like a film projector only in the sense that an image is projected on a screen. In a video projector, a movie is produced one picture at a time on film. A light shines behind the film strip; the image passes through a lens, from which it spreads out and is capable of covering a large area. Also, the lens can be turned to focus the image on the screen.

On the other hand, the video projector operates like a projection TV set, but with much greater distance between the screen and the

Fig. 2-13 Are all projectors the same? Be sure to test them and bring along plenty of money if you're serious. The VPH-1001Q (top) lists at $5990. Sony (top) and Runco (bottom)

light source. The *projector* is a large ceiling-mounted box (Fig. 2-13). Instead of a single light beam that shines through the film, this projector contains separate, powerful red, green, and blue lenses that beam the light on the screen, dot by dot (Fig. 2-14).

Fig. 2-14 A video screen, just waiting waiting for a herd of movie hounds with tubs of greasy popcorn to turn on the projector. Wee Bee Audio

It might seem impractical to beam the light to a wall or screen. After all, it seems that the resolution would be mediocre at best and the picture would be dark and "muddy." Evidently, these problems have been eliminated. A perfect example of the cutting edge of this technology is the Runco LightAmp 4000. Instead of displaying a mediocre NTSC picture, this box is compatible with HDTV standards and it can display at resolutions as high as 1200 lines! To be sure that the picture is strong and bright on the screen, a 1.5-kW xenon arc lamp beams the image forward. In fact, the Runco LightAmp 4000 is not a less-qualified technology that sells itself only on its eccentricities, but it is a high-end niche. I'd bet that this system looks better than any TV you've ever seen, but with a list price of $89,995, it should!

Cost is not the only major problem with video projectors. Not only are they more expensive than many houses currently on the market, but they require a large room, preferably with plenty of seating and space for one huge (in house terms) screen. Because of these prerequisites, practically the only people whose systems grace the pages of the home theater magazines seem to be those who are sports heroes, entertainers, or those in the movie/video industry.

COMPUTER VIDEO EQUIPMENT

Earlier in this chapter, a brief section covered the impending convergence of TV broadcasting and computerized Internet media. These dreams are presently somewhere off in the future, but the big question is when.

However, not every aspect of computer video is off in the distance. Right now, you can view plenty of video—some of which is at NTSC standards, and some that is at much better than those standards.

The first type of computer viewing is standard broadcast television via a PC card or an external box. A good example of this is the Hauppauge WinCast TV board. This PC card fits into the expansion slots on a home computer. After plugging the board into your computer, you would simply need to install the software before being able to use the TV. The final displayed TV picture would be similar to that on a standard NTSC TV, except that it would be noninterlaced, rather than interlaced. This is kind of handy if you spend hours at the computer and would like to keep track of news, sports scores, or a TV program. Really, the only advantage to using one of these TV cards is the convenience of watching TV on your computer while you work. Because of the time that less-attentive employees could spend watching TV, rather than working, these TV cards are not a favorite of employers!

If you are looking for a TV card for your computer, look at the description to be sure that you get the right card. A number of the cards and boxes out there are either TV adapters or screen grabbers. A TV adapter is a box that operates as an interface between your TV and your computer. It translates the TV signal into one that the computer can accept and then displays it on the monitor. This type of system might be useful if you have a home office or if you spend lots of time at the computer, but it is far less flexible.

Screen grabbers allow TV images to be displayed on a computer, where they can then be manipulated. These pieces of equipment have many practical uses—everything from using images of your kids from a video tape on a Web page to actually editing video digitally. Screen grabbers are often advertised as "putting TV video on your computer," so read carefully when looking for a TV board! For more information on video-capture cards, see Chapter 8.

WHAT TO CHOOSE

TV theories often seem very black and white, so to speak. Instead, everything becomes terribly complicated when you actually go out into the video marketplace to hunt down your next system. Most people's purchases are based primarily around the sales that turn up at the local appliance store, not on what TV would best suit their needs or tastes.

Your first priorities in purchasing a TV should be the combination of price and manufacturing quality/warranties. Essentially, this combination can be equated to the cost of owning the set per year that you use it. For example, my parents have been using the same Zenith console TV for more than 20 years and the performance has been consistently good. Divide the original purchase price by the number of years that it was owned and you get the cost per year. Some of today's TVs will be lucky to survive even five years. The difference in cost if you plan to keep your TV for many years is tremendous. For example, if you pay $279 for a TV that lasts for only six years, it will cost $46.50 per year to own. A $429 TV that lasts 17 years will only cost $25.24 per year, less than half the price of the "bargain" model.

Probably the best place to find information concerning the repair rates of TVs is *Consumer Reports*. This magazine regularly tests all sorts of products and surveys its readership. They rate products according to all sorts of subjective criteria, some of which I disagree with. So, don't choose a TV based solely on the ratings in *Consumer Reports*, but use their opinions as a springboard to formulate your own. However, one of the regular additions to most feature stories in *Consumer Reports* is a survey of readers concerning the repair rate of their product. For example, one of their most recent surveys listed more than 20 different TV manufacturers, from the lowest incidence of repairs to the highest. Of course, these ratings are not foolproof; you could get a dud TV from a manufacturer with a great reputation or vice versa. But, chances are that the ratings pretty much follow real life.

Now that you have an idea about the best TV to choose, it's time to take a look at the other features of the set. The best way to do this is to drop on by the local appliance/electronic store and have a look at their models. First, inquire about the warranties for the various manufacturers and take note of which are carried in that particular store. Most of these stores only sell video equipment from about five different companies. To view the best selection of equipment, be sure to visit at least a few stores before making a decision.

If you are satisfied with the warranties and the reputation of the company's TVs, then look further. Compare the sharpness and color to other comparable sets. Try out the various features on the remote control, such as picture in a picture, scan, volume and channel changing, closed captioning, etc. Except for the hearing-impaired viewer who would most likely rule out any TV that didn't include closed captioning, the features should be among the last quibbling points.

CONCLUSION

In the past, choosing the system that suited your video needs was simple—you just went out and bought the largest TV that you could afford. Now, it's much more difficult. Not only do you have to find a television or monitor that meets your criteria, but you must consider such issues as compatibility with future systems. Make the right decision and you could be happy for the life of the system; make the wrong choice and you could be Betamaxed!

3 VIDEOTAPE

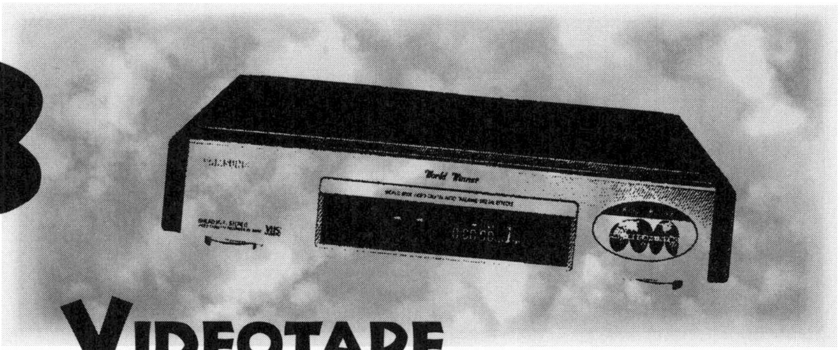

Like its sister technology, the standard audio cassette, the VHS cassette tape is the people's choice. Over the past 25 years, it has devastated any other video recording or playback technology that has dared to challenge it. That's not to say that it is the best technology for the purpose, but it certainly is the most widely used.

LIGHT VIDEOTAPE HISTORY

Like most forms of video technology, the history of most any video equipment goes back to a comparable piece of audio equipment. At one time not all that long ago, it was nearly impossible for most people to record and store any form of audio or video. You could do little more than listen to records or the radio. In the 1930s, some record-cutting equipment began to seep into the market. This wasn't record company equipment; it was actually intended for consumers. It wasn't terribly popular because the technology was very new and the Depression was still depleting the spare cash of most people.

During World War I, several countries began using wire recorders, with the wire magnetized, instead of the coating on a plastic tape that has been used for about the past 50 years. In fact, one of the first uses of the wire recorder was for espionage. According to an article by Tom Kneitel in *Popular Communications* magazine, the German government operated a radio communications station in the United States under the guise of a commercial station. The station recorded messages for its agents in the field on wire recorders. To fool everyone, the wire reels were played back at much faster speeds. For some time, no one could figure out what the high-pitched noise on the radio was (Remember, this was 1916!). Finally, someone with a new-fangled wire recorder recorded the transmission and decided to play around with the speed settings. Voila!

Of course, tape was out of the question in the Teens because most forms of plastic didn't even exist then. Through the 1920s and 1930s, some forms of plastic were developed and widely used. One of the most common was bakelite, which was used in radio cabinets and black telephones. Bakelite is hard and brittle; there's probably a better chance that you could make tape out of cheese than bakelite.

Everything changed during World War II. Vast amounts of money (spent by both the Axis and the Allies) were poured into research for all sorts of technological advancements. Naturally, plastics were extensively developed. During this time period, the open-reel tape deck was created and implemented, again by the Germans. By the post-war period, when the U.S. economy was ferocious, open-reel tape recorders hit the market and became an instant favorite. Unlike the competitive market that exists today, the electronics market of the late 1940s and 1950s was relatively open, and open-reel recorders quickly dominated.

With open-reel audio recorders so quickly in the bag, engineers from all of the major consumer electronics manufacturers threw their efforts into developing videotape recorders. The great push for this equipment came from the television industry, not from consumers. In the 1940s and 1950s, all television programs were either aired live on camera or they were played back from a film projector to a screen. Then, a special camera with a matching frames per second rate would capture the production. The problems were that film was very expensive and it delayed immediate retransmission, which prohibited most programming from being aired via this medium. But live programs were limited because the scripts had to be written to accommodate the few makeup or set changes that could be aired. Worse yet, mistakes could be costly to the TV stations and broadcasting networks—especially where the advertisers were concerned.

When the ever-splashy David Sarnoff was at the helm of RCA, he proudly displayed a videotape player/recorder in 1953. With RCA owning the largest TV network in the world at the time, the videotape recorder was of paramount importance. Unlike today's tiny cassette deck-sized consumer VCRs, Sarnoff's original display model approached eight feet in height and was six feet wide! Also, the unit didn't use a VHS cassette or even now-standard one-inch wide studio audiotape. Instead, it used standard 1/4-inch open-reel tape, cranked through at the whopping speed of 20 inches per second! Imagine how fast it would have had to travel in order to pass color information, too.

By the early 1960s, videotape had been established in the professional markets, and Sony began to develop a videotape player/recorder for consumers. This unit was very different from VCRs of any era, both technically and in appearance. Instead, this model was very similar to an open-reel tape deck. It proved to be very unpopular because of its high cost (around $2000) and inconvenience (the tape was held on spools that had to be hand-threaded onto the take-up reel).

The first order of business was to reduce the inconvenience of the system. Relying on the concept that made the audio cassette a hit, the manufacturers started working on a system whereby the tape would be enclosed in a manageably sized cassette-type shell. The first actual VCR was released in 1975, where it reached an enthusiastic, but small-spending audience. This battle was compounded in the early 1980s when the two video cassette formats, Beta and VHS, slugged it out in a close battle. During that time, many people were reluctant to purchase VCRs because they had no idea which format would rule the market and which would become obsolete (Fig. 3-1).

Since the mid-1980s, VHS has been the dominant VCR format although it has been getting pushed out of the camcorder market by

Fig. 3-1 An early advertisement for video equipment in the Beta format, which was finally driven from the North American marketplace about 10 years after it was introduced.

the smaller VHS-C and 8-mm formats. With the millions of VCRs that are already in homes across North America and the potential of recordable digital discs in the future, the VHS format might never be eclipsed by another videotape format.

THEORY

Before delving into the technology of VCRs, which can be fairly complicated, it's best to start with the technology that it was originally derived from, the open-reel tape deck. Open-reel (also known as *reel-to-reel*) tape decks use the same principles as cassette decks to produce sound. The sound is electromagnetically recorded onto the oxide coating of a plastic-based tape. This tape is fed from reels through a covered section that contains the tape heads, capstan, and pinch roller (Fig. 3-2). The only physical difference between cassette and open-reel formats is that with the open-reel format, the take-up and supply reels are separate and loose—there is no plastic shell to protect the tape. The particular formation of the magnetic particles on the tape induces a small

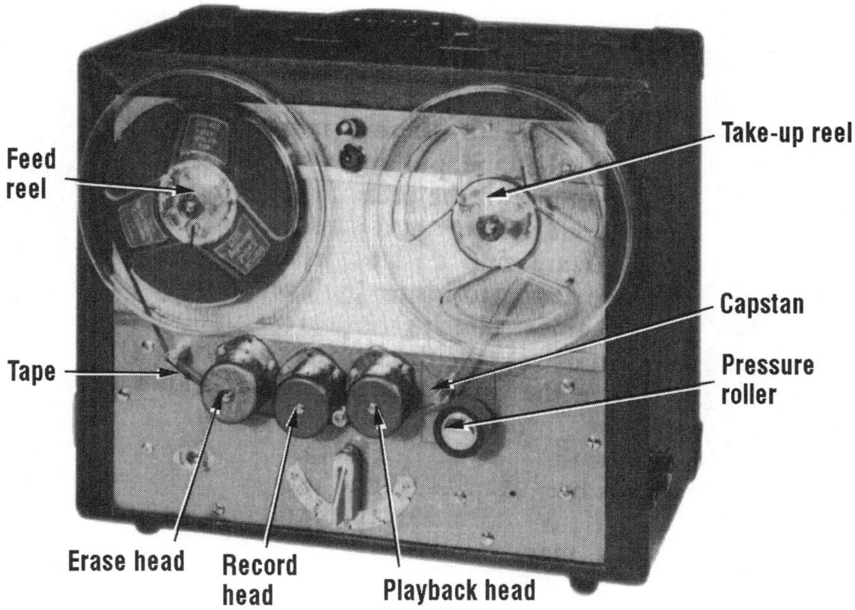

Fig. 3-2 A view of an older model of open-reel tape deck, showing all of the major components in the tape path.

signal voltage in the heads. These signals are then amplified and later turned back into audio by the speakers.

The differences between the two analog tape formats are minor in theory, but major when you actually use them. "Popping" a tape in a cassette deck requires less than five seconds. With an open-reel deck, you must pull the tape reel out of its box, place it on the supply reel pin, and pull the end of the loose tape around the capstan and the heads and up around to the pickup reel. Then you have to stick the end of the tape into the pickup reel in the same manner that you would load film into a camera. If the tape isn't placed in well, it will pull off the reel and spill all over the table or floor. Also, you have to fix the tape reels in place with end caps. Without them, the tape will fall off the pins and roll across the floor!

As you can see, just preparing to use an open-reel tape deck is much more difficult than using a cassette deck. Besides the difficulty of operation, most people get timid when they handle recording tape.

The early open-reel videotape players operated on virtually the same principles as open-reel audiotape decks. They looked almost exactly alike, with the same key components (heads, pickup reel, take-up reel, capstan, etc.) in the same general locations. The key difference was that the heads were divided into separate tracks, much like the stereo tracks of a cassette or open-reel deck. Instead of being divided into separate audio tracks, the open-reel videotape was divided into an audio track on one side of the tape and a video track on the other side. The heads were likewise divided into video and audio sections; the signals from each traveled off to their respective circuitry.

As mentioned, the manufacturers realized that videotape had to be convenient and easy to use or the number of users would be limited. To make it so, they would have to reduce the size of the tape so that it could fit into a manageable case. Taking a cue from the audio cassette, they developed a video cassette. The final videocassette player, although based on the openreel and audiocassette players in theory, is much more complicated.

First, take a look at a VHS videocassette. When comparing it to an audiocassette, the first difference that you would probably notice is that it's a whole lot bigger. In addition to the thicker, heavier case construction, the tape itself is 1/2" wide. This is more than twice the width of audiocassette tape.

Also, the engineers who specialized in magnetic recording discovered that the tape could be more efficiently filled with information if the tracks were recorded diagonally across its width (Fig. 3-3). Some of the different formats record at different angles, which are known as the *tracking angles*. The VHS VCR plays and records with a tracking angle of six degrees. The only problem with setting the information on a tracking angle is that it would then be impossible for the tape to properly play back by passing over a stationary head. This technique had numerous advantages over the

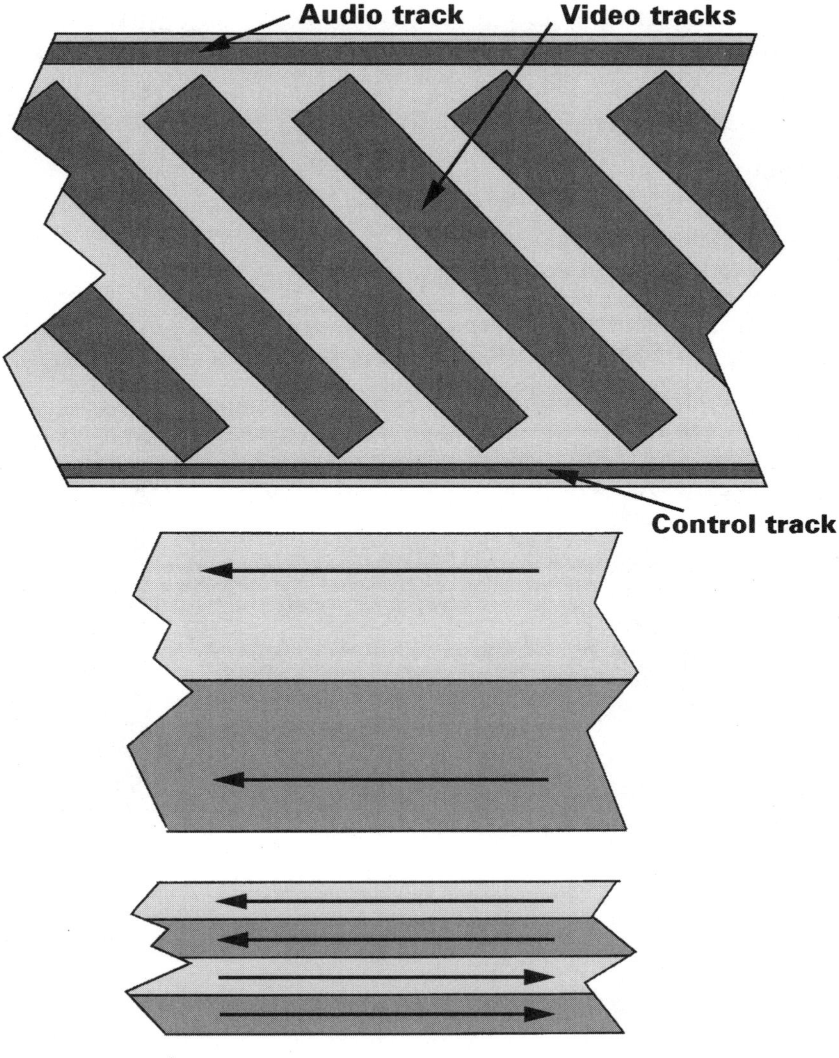

Fig. 3-3 A "larger than life" look at the tracks in VHS videotape (top), open-reel tape (middle), and standard cassette tape (bottom).

stationary head technique. For example, to pass such a large amount of information, the tape would need to be flowing by at an enormous rate of speed. At high speeds, the tape begins to pull away from the head, causing poor signal reproduction.

Because higher speeds require higher tension, it is possible for the tape to stretch or even break. To avoid higher speeds, yet have the same overall effect, spinning tape heads were necessary. With a bit of serendipity, spinning tape heads were required both to use a tracking angle (more efficiently using the tape) and to lower the necessary tape speed (preventing the tape from stretching). To better gather the video information from the tape, the spinning head drum was fitted with two video heads.

Despite the tracking angles that were used, the tape speeds were still high enough to cause the tape to pull away from the heads. To reduce the effects of tape separation, the tape is mechanically

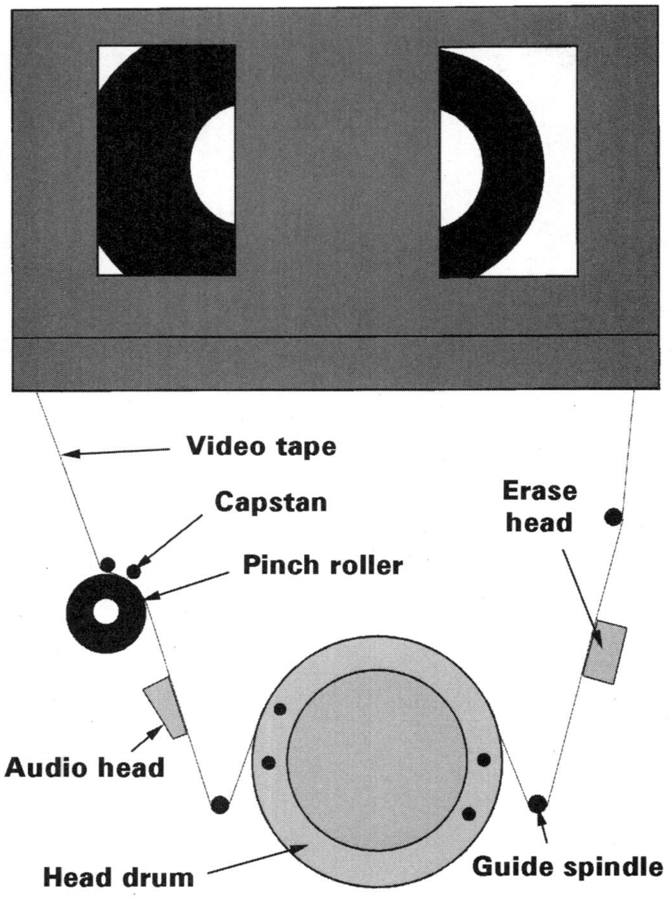

Fig. 3-4 The videotape is pulled from the cassette shell, against the stationary audio and erase heads, and around the spinning video head drum.

wrapped around the head drum (Fig. 3-4).
Unlike the video information, the audio is still contained on a linear trail along the one edge of the tape. It is tracked the same as that of a cassette or open-reel tape player, with a separate, stationary audio head reading one track (if mono) or two tracks (if stereo). On the other side of the videotape is another long, straight track of information. Instead of either audio or video, this track only contains timing information to keep the audio and video all in sync and running smoothly.

Originally, VHS VCRs were only capable of playing back and recording mono signals. For years, JVC engineers strived to obtain stereo audio. They finally achieved it by including stereo audio in with the angled video tracks. But, when they created this addition, the audio had to also be recorded onto the linear mono tracks or the older systems of VHS mono VCRs would be rendered obsolete. The "hi-fi" information is picked up by two audio heads that are added to opposite sides of the spinning video head. So, a typical four-head VCR has two video heads and two stereo audio heads. When a program is recorded on a four-head VCR, the audio information is saved both on the angled and the linear audio tracks.

THE TAPE AND HEADS

Originally, magnetic audio products used wire to record (as mentioned at the beginning of this chapter). Later, the magnetic dust was applied to paper tape. Soon after, the experimenters switched over to polyester tape coated with iron oxide particles.

The recording head consists of a coil of wire wrapped around a ferromagnetic "doughnut" element. A tiny open gap is in the bottom. This unit is placed within a metal case for strength and durability. When current passes through the coil, the magnetic field extends past the gap. The signals passing through the record head vary, depending on the volume and frequencies. These signals are then impressed on the tape. Peculiar as it might seem, the tiny particles on the tape can all be magnetically aligned, like microscopic polar magnets. The particles are brought into proper alignment by the induced magnetic fields from the heads. Because the audio and video depend entirely on the alignment of the particles, it is of utmost importance for the heads to be perfectly aligned and free of jitter.

Like the recording head, the playback head consists of a coil of wire wrapped around a ferromagnetic element. In many cases, the recording head and the playback head are both contained in a single unit. Basically, the opposite of the recording effect occurs. The magnetic field of the tape tries to pass through the gap in the heads—almost as if it is getting sucked in. As it passes through the coil, it causes a voltage to flow. This voltage is amplified, then is passed on to an amplifier, where the signal is increased in strength (amplitude). Then, it is passed on to the speakers, where the signal is reconstituted back into real-sounding audio (hopefully).

Nearly the same type of transformation occurs with the video signal. The magnetic field of the tape tries to pass through the gap in the heads—almost as if it is getting sucked in. As it passes through the coil, it causes a voltage to flow. This voltage is amplified, then is passed on to an amplifier, where the signal is increased in strength (amplitude). Then, it is divided into red, green, and blue components, which are shot one by one from the RGB guns to the front of the screen. If everything works properly, the video will be sharp and the color life-like.

VCRs ON THE MARKET

Times have changed since the days of the $2000 videotape player. Prices continue to drop, and it's now as cheap to pick up a decent VCR as it is to buy a good cassette deck. Just a few years ago, most four-head VCRs cost $250 and up, and most hi-fi VCRs were at $300+. Now, many hi-fi VCRs are selling for less than $200 and some mono VCPs have broken the $100 barrier! With more than 100 million VCRs in service and 12 million VCRs selling annually, this format is set to remain popular for at least another decade, if not longer.

With this relatively long time frame still available, you can comfortably purchase a new VCR, without the ominous specter of obsolescence looming nearby.

VHS IS VHS...OR IS IT?

Up until this point in the chapter, it has been assumed that VHS means VHS, except for the difference between hi-fi and mono VHS. Unlike the audio and video technologies of the past, manufacturers are discovering that they can manipulate all of that tape width.

The first video mutation from the standard VHS format was S-VHS (Super VHS). This format provides an improved picture from the VCR; in fact, the picture is approximately 30% more detailed than that from a standard VHS VCR. Best of all, the format is downwardly compatible—you can play standard VHS cassettes on an S-VHS VCR. But there are a few complications. The videocassettes that you record on your S-VHS VCR are not compatible with a regular VCR. Also, all of those movies that you can buy or rent from your local video store are recorded simply in the VHS format; playing them back on an S-VHS VCR will not enhance the performance. In the end, the S-VHS format is perfect for those who record from clear TV channels (cable, terrestrial, or satellite broadcasting) or those who record and edit S-VHS camcorder videocassettes. If your applications are something different from these, don't waste your money on the improved technology.

The next mutation to our favorite video format is just on the horizon at this writing. The D-VHS (Digital VHS) VCR is an ugly-duckling niche technology. This one can play or record in the standard VHS format, just like any nice VCR. But, with a flick of the switch (and the proper broadcasting service), the Clark Kent of the video world suddenly transforms into the Superman VCR (not to be confused with Super VHS). The D-VHS VCR (as you've probably caught on by now) can also record and play back digital signals with fantastic results (Fig. 3-5). This digital monster can reproduce video at a horizontal resolution of up to 500 lines. But there are a few catches. You can only play back your signals through a digital satellite receiver. Worse yet, there has been some speculation about whether one manufacturer's D-VHS VCR will be

Fig. 3-5 The digital (D-VHS) VCR--a fantastic new development that will impact all future video designs or another white elephant in the vast world of technology? Sony

compatible with another's satellite receiver. The result is messy: The format can essentially only be used by those who have digital satellite service. If there are receiver-compatibility problems, the potential for this format is even further limited.

The D-VHS format seems to be doomed from the beginning, just like its digital audio sister, the Digital Compact Cassette (DCC), which failed miserably in 1994. Perhaps, if HDTV can make it off the ground, a modified version of D-VHS will be successful. If so, this format would be able to record straight from HDTV broadcasts and play back on HDTVs, yet still be able to record and playback in the standard analog VHS NTSC specifications. This type of transition will be necessary if HDTV is ever to become successful.

OTHER FORMATS

If you have shopped for VCRs for any length of time, you will realize that standards other than the VHS variations and ancient Beta decks are available. But what are these beasts and why would you want to use them?

By far, the most common of these decks is the 8-mm VCR. The 8-mm tape looks very similar to a standard cassette. In fact, I've gone into some department stores and mistakenly grabbed a pack of 8-mm tapes instead of audiocassettes.

The big question is why would anyone depart from the VHS standards. For typical playback and recording, there's no reason to go with an 8-mm VCR. You're not going to be able to find movies in the 8-mm format. The whole reason for the existence of 8-mm VCRs is camcorders. The 8-mm format was developed specifically so that camcorders could be made lighter and easier to carry. It caught on because people were tired of lugging around full-sized VHS camcorders and having various body parts fall asleep from the weight. Likewise, people with many, many home videos (recorded from their camcorder) could play them back on their own 8-mm VCRs.

An 8-mm VCR should only be of interest to you if you already own an 8-mm camcorder and you have loads of 8-mm tapes. Also, you can purchase double-deck VCRs that contain one 8-mm and one VHS deck. The intention here is that you can record from your 8-mm deck to a VHS tape (or vice versa). This system might sound handy, but if you already have an 8-mm camcorder and a VHS VCR, you can save lots of money just by recording between them. Finally, if anyone invites you over to watch videos from their 8-mm VCR, be very wary! For more information on camcorders, see the second half of this chapter.

Now, a Hi8 VCR is another story. The Hi8 format is essentially an improved version of the 8-mm format, much the same as the being S-VHS is an improved version of VHS. Rather than just a toy for home movie fanatics, the Hi8 VCRs are specifically intended for high-quality recorded broadcasts, and for amateur and professional video editing applications (Fig. 3-6). With Hi8 VCRs priced anywhere from under $1000 to more than $7000, this format is not just a miniaturized novelty version of a "real" format.

FEATURES

hi-fi As mentioned in the text, hi-fi VCRs simply have stereo audio. This is a real boost in realism when using such a unit with a home theater system. Considering the difference in prices between VCRs with or without this feature and the improved sound, a hi-fi VCR is a must.

Fig. 3-6 Looking for better performance from a videotape? Maybe you should look into Hi-8 VCR. Sony

four head and six head Four-head VCRs contain four different heads spinning on the head assembly. You can generally assume that a four-head VCR will offer superior performance to a two-head model. However, new two-head VCRs are becoming increasingly rare anyway. Expect that six-head VCRs will provide even better performance, particularly at slow speeds.

on-screen display Most of the new VCRs offer on-screen display. With this function, you can press a button and some text programming will appear on the TV screen. Then, you can pop a menu screen up on the TV, press a few Up, Down, and Enter buttons on the remote control, and voila!, the VCR is programmed. If you are hiding illegal aliens in your household, they will be pleased if your VCR has a multi-lingual display. Most VCRs sold in North America contain English, Spanish, or French.

remote control Like the aforementioned features, remote control is available on nearly every VCR. Considering that VCR tuners are often used to control cable channels, it would be an incredible inconvenience not to have a VCR with remote control.

timer VCR timers are yet another necessity of the modern communications age. Now, you don't have to miss badminton tournament night so that you can stay home watching the latest episode of "Hard Copy." Just set the timer and the VCR will record the program, starting and ending at the times that you programmed. You can even watch a program on one channel while recording something from another.

cable-ready tuner VCRs are equipped with a tuning section so that you can control the cable TV channels with the VCR remote control. The only feature that you need to consider when looking at the VCR tuners is whether it can cover enough channels. Most cover about 180+ channels, so this is rarely a problem.

three-speed playback and record (SP/LP/EP) Nearly all VCRs can play or record at SP, LP, or EP. The difference between these settings is the speed at which the tape is played back. A typical T-120 videotape will play for 120 minutes in SP mode, 240 minutes in LP mode, or 360 minutes in EP mode. In the end, you need to determine whether you want to sacrifice quality of picture for increased recording time.

counter Most VCRs feature a tape counter of some sort. VCRs from the 1970s and 1980s used mechanical counters, but now digital counters are used exclusively. The best counters track in real time, not just four digits, set at some random increment.

tracking In days past, the tracking was always controlled (-/+) by a thumbwheel located in the front panel of the VCR. Now, most tracking controls are operated digitally from the remote. Instead of being on an analog knob, tracking is controlled by clicking the + or the - button, so the tracking is altered in steps. Some of the feature-laden VCRs also include auto tracking. Personally, I've had problems with all electronic equipment that controlled anything automatically, so I'd avoid this convenience.

auto head cleaner Many manufactures have discovered that the people who purchase their VCRs don't really take care of them. In fact, they don't maintain them at all. To help save the VCRs of the complacent, some companies are using automatic head cleaners. Anytime you pop in a new tape, the heads get cleaned. It's kind of handy, but nothing that some alcohol and a cotton swab stick couldn't handle.

picture sharpness controls Many VCRs presently available have some sort of proprietary sharpness, clarity, or color circuitry. These are sold under different names, such as HQ, and they perform different tasks. Rather than just buying a VCR on the catchy-sounding names for the circuitry, take a look at the VCR's performance for yourself.

slow motion This function controls the speed of the videotape playback. As anyone used to watching televised sports knows, this is the ability to play back a tape at a slower speed.

input and output jacks Most new VCRs contain input and output jacks on the front panel. This makes is much easier to quickly dub or edit videos from another VCR or from a camcorder (Fig. 3-7).

Fig. 3-7 A close-up of a jack panel, showing the video input, stereo audio inputs, and S-VHS input. Sony

AC/DC Have you ever wished that you could connect your VCR to a car battery and watch movies anywhere? It's easy with an AC/DC VCR. One of these units contains power supplies so that the VCR can operate via either AC or DC power. These units are often combined with a television, for extra convenience. A must for "off the grid" homesteaders and Winnebago warriors alike.

visual search With visual search, you can play the videotape at a faster-than-normal speed. Then, you can visually search the tape for whatever scene you want to find. Some manufacturers call this double-speed playback.

freeze frame Freeze frame allows you to hit Pause, and the picture will freeze on a particular video frame. On VCRs without this feature, if you hit Pause, the video would typically stop between frames and look all blurry and distorted. A good freeze frame can be handy—especially if you want to know whether the running back *really* crossed the goal line or if you *did* see a jet in the background of that spaghetti western.

VCR+ VCR+ is a simplified on-screen programming system.

Commercial Advance Are you compelled to buy everything that pops up on TV? Do you find small distractions unbelievably annoying? If so, you will wonder what your life was like before Commercial Advance was on your VCR. This feature electronically marks the beginning and ending of commercials when you are recording programming. Then, when you play the tape back, the VCR will take note of the mark and fast forward the tape until it reaches the ending marker. Of course, if you would have used this feature to tape a Buffalo Bills Super Bowl game, you would have missed the best parts.

flying erase head If you edit video segments, then a VCR with a flying erase head is a must. A typical VCR will leave a few seconds of snowy/color-barred video at every edit. But if the deck has a flying erase head, the transition will be smooth. Only semi-professional and professional VCRs contain flying erase heads, although this feature is common in camcorders.

jog/shuttle control Like the flying erase head, the jog/shuttle control is a feature intended primarily for semi-pro or professional

VCR editing decks. The shuttle control is a knob that allows you to smoothly vary the search forward or reverse speed. With the jog control, you can advance the picture frame by frame. Beware of VCRs that have their controls aligned in a circle—from a distance or a small photo in a catalog, these VCRs look deceptively as if they include the jog/shuttle control feature (Fig. 3-8).

Fig. 3-8 Don't be fooled by all of the VCRs on the market with circularly aligned controls. Here's the real thing--a Samsung VCR with jog/shuttle knob for ultimate speed and editing control. Samsung

Index search This feature is somewhat like commercial advance. Whenever you start to record something, the VCR will place a marking signal. When you play back the tape, you can search for these index marks. When you combine these marks with a written list of titles on your tape, it's easy to search for programs.

sound alert Some VCRs will beep at different times for various functions. Do you really want your VCR to talk back to you?

child lock This function locks the VCR so that it cannot be operated until it is unlocked. Having children who like to play with technology and have already broken two of my cassette decks, I understand the usefulness of this measure!

XDS capability With XDS capability, your VCR will be capable of displaying current program information and the station ID of the channel you are watching. In effect, it is a TV version of radio's RDS, which would use an LED display to show the station's callsign and some other information.

blank search Just in case you don't bother to cue videotapes to the proper position before recording, blank search was created. With it, the VCR fast forwards through unrecorded (blank) tape until it finds something that has been recorded.

power backup The power backup feature means that a tiny amount of power is contained within the VCR circuitry when it is unplugged or when there is a power blackout. This power will maintain the programming memory within the VCR ICs until external power is once again restored. Most VCRs can only tolerate a shutdown of a few minutes, so if your cat dives behind your TV cabinet and your unit remains unplugged for two weeks while you're on vacation, forget it.

auto tape speed adjustment If you put in a video cassette that has been recorded at EP speed, the VCR will automatically detect the recording speed and match it. If you're buying a new VCR, this shouldn't even be a consideration.

speed watch This feature is just too great. One Fisher VCR features "speed watch," which allows you to watch videos at double speed (for busy people) and seven-times speed (for busy people who are either lacking attention spans or who are taking a bit too much medication). Best of all, the sound is electronically pitch corrected to sound normal. Imagine watching a 60-minute "X-Files" re-run in only 30 minutes...or better yet, in just over eight minutes (in 7X speed). "Wait! Was Scully abducted by huge, yellow, doughnut-eating aliens or was that just an ad for The Simpsons? Who cares, it'll be over in another 45 seconds anyway."

"Universal" VCRs Some people call these units "universal" VCRs, but their key function is to play back VHS videotapes in either the NTSC or the PAL video formats. One common problem for travelers is that they buy a videotape or trade one with someone in Europe. When they get the tape back, it doesn't play properly because it was recorded in the PAL format. Europe and Africa are split between the SECAM and PAL formats; the countries in Asia are split between NTSC, PAL, and SECAM; South America is

divided between NTSC and PAL; and North America is strictly NTSC. A universal VCR can play PAL, SECAM, or NTSC tapes, so you can take the VCR anywhere or watch VHS videotapes from anyone in the world (Fig. 3-9).

Fig. 3-9 The perfect VCR for the world traveler: the NTSC/SECAM/PAL compatible Samsung SV300W. This unit also includes various AC power supplies so that it can be operated most anywhere without any power converters. Samsung

PURCHASING A VCR

When it finally comes time to plunk down a load of cash for that VCR, what will you buy? Will it be the $99 generic VCP, the $8000 editing VCR with digital effects, or something else entirely? Chances are that you will want something in between these two models (Fig. 3-10).

Of all of the features listed, only a few are crucial to your decision. If you need a VCR for standard movie-viewing purposes, go for the following criteria:

* Hi-fi
* Four or more heads
* "Reliable" brand name
* Heavier-weight models

The last two of these points need at least a little explanation. A "reliable" brand name means one with fewer reports of repairs. For more information on this statistic, see VCR reviews in *Consumer Reports*. The last point is also mentioned in Chapter 1: You can

Fig. 3-10 Unlike the "universal" mode VCR, the Samsung VR5607 is more of an "everyman's" model. Samsung

often justifiably assume that a heavier piece of electronic equipment is made better than a lighter model (assuming that both are new). This isn't necessarily true for the more expensive models, but you can bet that a cheap model won't be heavy.

Aside from these points, look for others of importance. For example, if you are planing to travel frequently in a camper, an AC/DC VCR would be nearly essential for the trip. Other key features would include a flying erase head and jog/shuttle control for those who plan to edit on their VCRs. Aside from the importance of these features, almost all of the others are just icing. It's best to forget about most of these and focus on the quality of the machine.

CAMCORDERS

Perhaps the single greatest contribution to the preservation of personal and family history in the twentieth century (arguably, forever) has been the development of the camcorder. With it, you can display live-action video and audio. You can experience the birth of a child, a wedding or birthday party, or the time that you tried to fly off the roof with feathers glued onto your arms.

BACKGROUND

You might think that the camcorder traces its lineage back through commercial reporter-grade video cameras to the massive early TV cameras. In a sense, it does, but the most-direct connection (at least in terms of features, operation, and styling) is to the old 8-mm and

16-mm film cameras. These units were light, self-contained, and relatively inexpensive. The most expensive aspect of creating your own home movies with this technology was developing this film. If you are a bit too young to have experienced home movies on film, these cameras were very similar to real photographic cameras; the film would move through the camera and picture after picture would be exposed on the film. When you were finished, you would have to send the roll to a photo developer, who would create a roll of movie film: a long reel of slide-type pictures on a clear film. When you received the reel back from the printer, you could pop it in your film projector, which would shine a light behind the moving film and shine the moving images on your huge projector screen.

This all sounds very archaic, but the quality could be excellent with a good 16-mm camera and projector. The problems were all tied up in the cost and inconvenience of the medium. It was much more expensive to develop a movie reel than a roll of pictures, so most people only shot movies on very special occasions, such as weddings, birthday parties, or Christmas celebrations. Editing and special effects were next to impossible, projector screens were bulky and inconvenient to haul around, and forget about recording any TV programs or movies!

With the many difficulties of video film making, consumer videotape recording was a given. The problem from the beginning was just miniaturizing the technology so that you could carry it (let alone comfortably haul it!). With the VHS and Beta formats being created in the late 1970s, videotape cameras made little impact until the early 1980s. Unlike today's camcorders, some of which are about the size of a half-gallon jug, early models came in two separate pieces—the video camera and the portable VCR. To shoot some video, you would pop a blank tape and a fully charged battery pack into the VCR and hang it by your waist with a shoulder strap. Then you could pull out the video camera and plug it into the VCR. The equipment worked well, but was neither convenient nor easy to use. Although the VCR was held at your center of gravity, a long videotaping session could really wear you out, which is particularly bad when trying to hold a video camera smooth and steady (and smooth, steady camera work is the key to good videotaping).

By the mid-1980s, many video cameras were manufactured with the VCR section right in the camera unit. These new cameras were called *camcorders*, and they were a vast improvement. The one problem with the new camcorders was that the videotape was too large for the camera to be made really small or lightweight. Instead, you had to lug a big, heavy camera on your shoulder for hours at a time. It's no wonder that during this era of the camcorder, about the only people that you ever saw professionally recording events were big, sturdy guys!

The late 1980s were a time of innovation in the consumer video world: the Beta format was finally put to rest in North America, and the VHS-C, S-VHS, and 8-mm formats were introduced. VHS-C and 8 mm were particularly helpful to the home videotaper because these miniature tape formats meant smaller camcorders. At last, you could comfortably record events without having the shoulders of a linebacker! Since this time, there have been relatively few innovations that affect the major qualities of the camcorder. The most drastic innovations have involved dropping the prices, further miniaturizing the parts, automating more aspects of the recording process, and improving the quality of the audio and thevideo.

FEATURES

A number of different features are incorporated into camcorders. Most of these use proprietary circuitry, so it seems like there are different names for about half of the similar features on the market. As a result, the features in this section aren't complete and some of these are included in other camcorders under different names. Still, this section will provide you with plenty of information to get you started when the salesman starts tossing out a barrage of high-tech names. See Fig. 3-11 for a variety of camcorders.

input and output jacks These jacks are just as important on your camcorder as on your VCR. Without these jacks, you can't record video directly to a VCR, which is essential unless you have several

Fig. 3-11 A variety of camcorders, showing some of the different features and styles. Sony (top) and Samsung (bottom)

VCRs and don't mind dubbing there.

auto focus Why bother to even touch the camera? With auto focus, just hit record and point the camera at anything; the camera will do the rest.

auto white balance Are you sick of carrying that big white piece of paper around so that you can white balance the camera before

Fig. 3-11 (Continued) More varieties of camcorders. Sony (top) and Samsung (bottom)

shooting every time? The auto white balance feature sets the color properly so that your shades of yellow don't wind up looking medium green, etc.

color viewfinder With the color viewfinder, you can see what you're recording in color, rather than in black and white. It's not a great advantage, but it might help you catch some video bits that you would otherwise miss.

tripod compatibility This feature makes the camcorder compatible with the fittings on any tripod so that you can smoothly shoot weddings or some other event, even if you forget your own tripod.

auto audio record level With the automatic audio record level, the camcorder will automatically set the level of the audio so that it isn't too loud or too quiet. Anywhere else, this would be called a compressor/limiter.

date/time recording Most camcorders will automatically post the date and time on the screen when you shoot video. Press a button and you can remove the information. This one is really annoying when people leave the information posted on the screen for all of their videotaping.

improved-quality formats Here are great features for the professional videographer: formats that allow higher-quality images than are possible with standard VHS videotape. Someof the leaders in sales are the S-VHS and 8-mm formats, although theHi-8, and digital formats are on the rise (Fig. 3-12). These formats are primarily intended for professional or semiprofessional use (except for 8 mm) because they aren't compatible with the VHS standard. If you want to play the tapes back at high quality, you must either purchase a VCR in that same format or record the video onto a standard videotape (and lose video quality). For more information on these formats, see page 90).

fade Push a button and the screen will either fade to black or fade up from black. Fades really help smooth out the transitions in the video and make it more pleasant to watch.

flying erase head The flying erase head is essential for clean cuts when shooting video. A typical VCR will leave a few seconds of snowy/color-barred video at every edit. But if the deck has a flying erase head, the transition will be smooth.

Fig. 3-12 If high-quality audio and video are a must, look into the Sony Digital Handycam. Sony

rechargeable battery All camcorders have rechargeable batteries, but before making a purchase, consider how many the camcorder comes with, how long they last, and how much extra batteries cost. Camcorder batteries are typically quite expensive.

ac power supply All camcorders should come with an AC power supply, which doubles as a recharger.

external video screens Sometimes people have problems looking through that tiny viewfinder for long periods of time. Enter the external viewfinder. This little LCD screen allows you to watch the video that you are shooting from a comfortable distance. Personally, I am less distracted when using the viewfinder, but plenty of people seem to like the external screens better.

recording indicator Most video cameras have two recording indicators—one in the viewfinder and one on the outside of the camera, above the lens. This allows both you and the people that you are taping to know that you are, in fact, taping them. This feature is handy, unless you are undercover, shooting footage for illegal government sting operations.

playback compatibility As mentioned earlier in the chapter, the compatibility of the camcorder with your VCR can make a great difference in either the convenience of your viewing and taping or in the total cost of your system.

picture effects These days, most camcorders have at least a few special video effects, such as fade to or from black (fades a picture either up from black or fades it to an all-black picture), split screen (which divides the screen in half, with a mirrored image on the other side), and stop frame (which "takes" one frame and holds it for a predetermined amount of time, then "takes" another frame. The result is a jerky effect that is somewhat similar to the effects of a strobe light). Sometimes these effects are placed on different buttons, but to save valuable space on the side of the camcorder, they are often all incorporated into one button (Fig. 3-13).

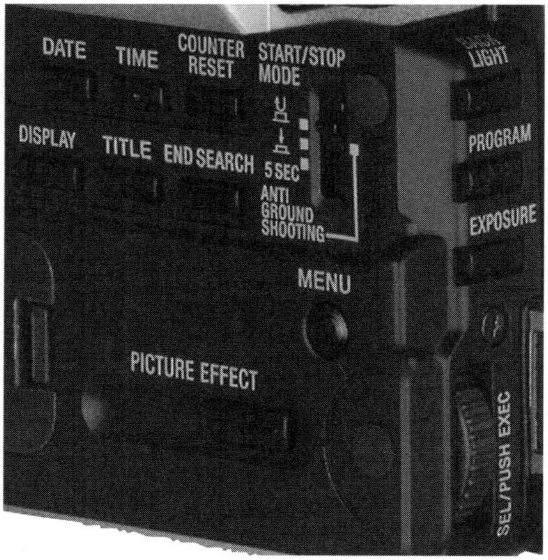

Fig. 3-13 Some of the many features that are included on the side of a Sony camcorder. Sony

AC power supplies and battery chargers An AC adapter/battery charger isn't really a feature per se, but it could significantly impact how, when, and where you videotape. For example, can it charge batteries and power the camera at the same time? How long is the cord for the unit? Are the battery charger and AC power supply combined in one unit or are they separate? Is the camera sold with both the battery charger and the AC power supply or do they cost extra? For a photo of a fairly typical AC power supply/battery charger unit, see Fig. 3-14).

Fig. 3-14 The ACV700 AC power adapter/battery charger . Sony

Special camcorders Camcorders are sometimes manufactured for special recording applications. Probably about the first special-purpose video cameras were the gun cameras that were installed in airplanes during World War II to record the "kills." Right now, the major application available is for recording footage underwater. Such a camcorder could be used for swimming classes, underwater animals or environment recording, boating, or just simply for videotaping during inclement weather. The result of this work is Sony's Marine Pack Handycam (Fig. 3-15). How long will it be before they develop the subminiature Surviellancecam?

LOOKING AT THE FEATURES

Three of the most important camcorder features aren't even true features: weight, length of operation per battery, and tape format.

Fig. 3-15 In a bright yellow case, just like the all-weather Walkman, here's the Marine Pack Handycam. Sony

The weight of a camcorder is a key factor—especially when choosing a used camcorder. Many of the 1980s camcorders are quite heavy. After 10 or 20 minutes, your arm would start to get the shakes from trying to hold the beast steady. Today, weight isn't as much of a problem, but it can still be a factor. In fact, some strong people also have a problem with camcorders that are too light; they are so light that the camera is jumpy because it doesn't provide enough resistance to be stabilized.

How long the battery will last on average while it is running is another consideration. This factor will make the difference between whether or not you need to buy a few more of the rechargeable batteries so that you have enough power to shoot an event. And these batteries are not cheap. They vary, but some are priced upwards of $50 apiece. Also, take the length-of-battery-life figure very liberally. Most are taken with the camera running in Standby mode. Actual camera operation will eat up plenty more power. Assume that actual running battery time will be about half of the Standby operating time.

The tape format can be very important for a few different reasons. One of the most obvious is that a small tape format means that the camcorder will be a lighter load to drag around (see the first paragraph in this section). Likewise, running a different tape format means that you will have some difficulties with compatibility, particularly if you plan to be editing a lot of video. Also, VHS videocassettes are much less expensive than VHS-C or 8-mm tapes. If you plan to record many tapes, this cost could significantly affect the total amount of money that you would spend on the system. If high-quality video is essential, then you need an S-VHS or Hi8 camcorder to go along with your compatible TV and VCR. If small size is essential, go with either VHS-C or 8 mm. If a low price is the bottom line, standard VHS is typically the cheapest medium in the long run (including the price of videotapes).

One feature can be more of a liability than a boon. Some low-end cameras have autofocus capability, but cannot be focused manually. This can cause the picture to stay out of focus for long periods of time, to focus on the wrong objects, or to not focus smoothly. Another "feature" that I find really annoying is electronic zoom. Most of the time, the zoom has hard starts and stops and is too fast. Not only is it annoying for these reasons, but most people seem to use it excessively—a motion-sickness nightmare. I guess this isn't the fault of the manufacturers, but it re-emphasizes to me that this is a bad thing.

Buying a camcorder

With a little background information on the available camcorder features, you can look around at video stores. Buying a camcorder is a bit different than buying a TV or a general-use VCR. Although you have to use these products regularly, the most important factors are the quality of the picture or signal presented and the reliability of the equipment. Camcorders are totally different. You have to be able to personally interact with the camcorder on a regular basis. If the camcorder is a bit uncomfortable or the buttons simply don't "feel right," it could hinder your videotaking abilities or cause you to miss great shots. In addition, the camcorder must also capably produce crisp pictures and accurate colors, have good battery performance, good zooms, and more.

If the camcorder's place in your life will be limited—for example, if you only plan to pull the dusty camcorder case off the shelf to catch a birthday party or two every year or maybe to tape your high school or family reunion every five years, then you probably don't care much about the features, quality, or ergonomics. If these details fit your situation, don't even bother with most of this chapter. Just get something cheap. Better yet, look for a used camcorder in the newspaper classifieds and purchase a low-priced model that works well.

On the other hand, if you plan to use the camcorder regularly, whether for personal or professional use, you have to "play" with the equipment before you buy. First, you should check some of the prices between the local electronics and appliance stores and the mail-order electronics outlets. You should have some idea of what you might want to purchase before you go into the store. For a better idea of what's available on the VCR and camcorder market, visit the newsstand or library and pick up a few issues of *Camcorder Buyer's Guide* and *Videomaker*.

By now, you have your list narrowed down to a handful (or two) of camcorders. Although you might be leaning toward a few of these units, you really have no way of knowing which one is best. There is only one way that you can accurately evaluate all of these features. You must actually try different camcorders at the store.

Some store managers might not want you to try out the camcorders, but most shouldn't complain too much if they want you to buy something. If they are more concerned with avoiding work than helping you, you're better off going to a different store anyway.

Finally, take in your own VHS, VHS-C, and/or 8-mm videotapes to record on, or buy them in the store. If you think that a better S-VHS or Hi8 camcorder is more your speed, take along some tapes of those formats, too. Be sure to let a salesperson know that you want to try out some camcorders, get his or her permission, then go for it. Pop the tape in and start recording. Try out all of the features and pay particular attention to the zoom.

Not only is testing out camcorders kind of fun, but it also makes for a great date as you and your significant other record each other making faces at the camera, waving at the salesperson, and dancing along with music from the audio systems. Juvenile? Certainly, but chances are that these situations are the types that the camcorders will be used for in real life anyway! Just be sure that you don't act clumsily with the camcorder—you break it, you bought it.

After several minutes of this fun, pop out the tape and try it on a good VCR connected to a good TV. The store should have some demo VCRs connected to TVs for evaluation purposes. If not, ask if the salesperson will hook one up. It's doubtful, but they might want to see the videotape enough to help out. Now, see how it looks. Unless you've had plenty of experience with videotaping, the final recording will look much different on the full-sized, color screen than when you were doing the taping. Notice the sharpness of the picture, the color, and the smoothness of the zoom. Try this with a few different cameras and see which you liked the best. If you are really serious about this stuff, go home and sleep on it. You can then make a clear decision about which camcorder offered you the best quality and ease of use for the lowest price.

CONCLUSION

VCRs and camcorders are the most popular pieces of equipment for storing video information. At present, both types of equipment have a virtual monopoly in the video world. But this is all going to change when other forms of digital media are developed. For more information on digital discs for video, see Chapter 4.

4 Digital discs

As mentioned in Chapter 2, the present conquering force in home video systems is analog videotape, particularly that in the VHS format. But the future is entirely different. As has been proclaimed from the Internet to the audio world, the future is digital. Another difference is that the dominant digital format will not be carried on some version of videotape.

As mentioned in Chapter 2, the D-VHS format is about to hit the market at this writing. One main disadvantage of D-VHS is that it's on tape, so it can be damaged from magnetic fields (such as from speakers or magnets themselves). Also, the digital disc players can immediately jump to any given scene or time in the video, but D-VHS tapes must be slowly rewound or fast forwarded. The big advantage of D-VHS is its compatibility with the analog VHS format. Many people will go with D-VHS simply because they will be able to continue playing their old collection of VHS movies, TV recordings, and home videos. The digital disc formats don't have this advantage, but they have many others.

Basics

Like understanding the technology behind video techniques (the section started with audiotape and worked back to video), it is easiest just to start this section with audio CDs. The compact disc is simply a thin disc of aluminum that is encased in a plastic laminate to protect the recording (Fig. 4-1). The music that is to be recorded onto compact discs must be in a digital medium; that is, it must be converted into a massive amount of 0s and 1s. When the disc is recorded, a multitude of error-correcting data and system information (track information, markers, etc.) are also added to the disc along with the music. All of this data must be downloaded, so the aluminum disc is etched with minuscule pits. The pits and the unpitted areas translate as data that represents the 1s and 0s.

One of the key components in a compact disc player is the laser optical assembly. When a compact disc is running inside of the head unit or changer, a low-powered laser is firing straight up through the player and at the tracks of the disc. The unpitted areas of the disc reflect the light back, but the pitted areas reflect almost nothing back. The result is a tremendously fast flickering of light—coded messages sent in the same manner, but much faster than could be signaled between the battleships in the old naval war movies. These coded reflections are received by a photodetector, which changes the light flickers into electrical impulses. These binary electrical impulses are then converted into analog impulses. Then, they are amplified and converted from electrical analog impulses into sound by the speakers.

Of course, this information about how a compact disc player operates is tremendously simplified. The analog-to-digital and digital-to-analog processes are extremely complicated—especially when you consider that such things as coding and sampling must also be configured into the system. *Sampling* is the process by which the compact disc player plays an analog sound, then checks the digital source, then plays another sound. This cycling occurs 44,100 times per second (44.1 kHz), although many players now sample several times more than that per second to make sure that the information being received or played is accurate and not error-ridden. Sampling at harmonic frequencies (as just described) is known as oversampling. Many of the high-cost compact disc players sample up to eight times the standard sample frequency. At this point, oversampling starts to become overkill and the law of diminishing returns kicks in.

Another factor, relative volume, should also be considered. Every

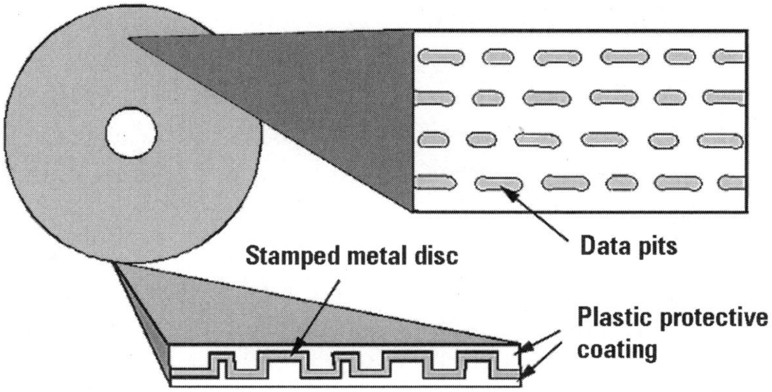

Fig. 4-1 Two super-close-up views of the CD: the surface pits and the side of the disc.

audio waveform has a length (the frequency of the sound), which determines the pitch of the sound and a height (the amplitude of the sound), which determines the volume of the sound. In order for the compact disc player to accurately reproduce music and not wind up with all of the frequencies reproduced at the same volume, the samples are quantized to a 16-bit number that lies between 0 and 65,535. Every tiny piece of audio that is reproduced by the compact disc can be reproduced at any one of the 65,536 different volume levels (Fig. 4-2)!

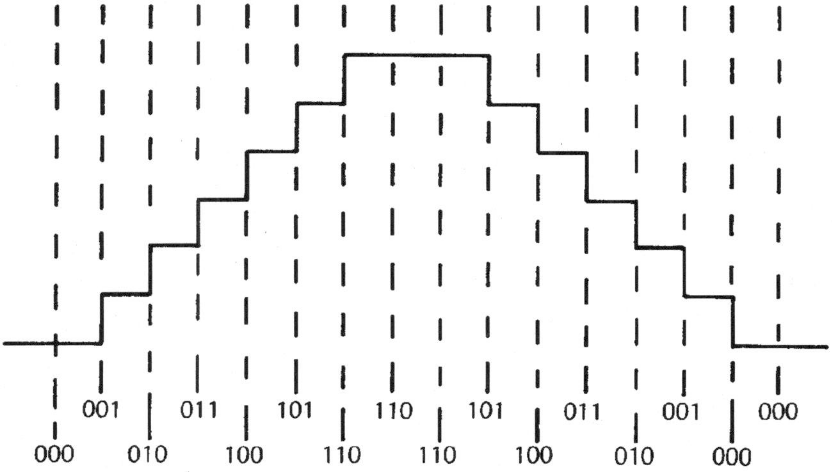

Fig. 4-2 A very rough representation of a digital audio wave.

These codes that determine various aspects of the compact disc's sound and technical operations all require a vast amount of information. A full compact disc (approximately 74 minutes) requires in the neighborhood of 34,000,000 bits of information to produce. If this information was all held on a standard computer floppy disc, the selection would have to be placed on 48 5.25-inch (13.34 cm) discs! The alternative to this is to determine some sort of compression code, which is how DAT and DCC tapes and MiniDiscs can be digital and hold as much music as they do.

THE LASERDISC

The laserdisc was the first real digital breakthough in home video systems, if you can necessarily call it a digital system. Unlike the typical CD format, the laserdisc contains several different audio tracks. In most cases, there are four separate tracks—two digital audio tracks (left and right) for the movie soundtrack and two analog audio tracks. The digital audio tracks reproduce audio just like audio CDs. The analog tracks can be any one of a number of different combinations. They could contain the movie soundtrack on both analog channels (left and right), a commentary about the movie running on one track and the movie soundtrack running on the other (both in mono), or a Dolby audio signal on one track and a regular mono audio track on the other.

The weird thing about laserdiscs is that they aren't entirely digital (as noted). After more than a decade of digital audio, you might have guessed that the discs would be entirely digital. Not so. Instead, two of the audio tracks and all of the video are in an FM-synthesized analog format. This system is based around the FM audio broadcast system that was described in Chapter 2. Rather than have a system of pits and lands to represent certain codes (more on this in the next paragraph), the laserdisc uses the system whereby a carrier is generated. The program information is determined by the frequency changes in the carrier (deviation) and by the distances between those shifts (phase). The greater the deviation, the louder the signal; the smaller the deviation, the higher the pitch.

The pits and lands of the videodisc represent the frequency variations of the signal, and the carrier signal is also recorded to the disc. In this sense, the video and "analog" audio signals are kind of a quasi-analog format—the analog waves are represented by digital-type pits and lands. So, when the laser reads these codes, the carrier is real and the pits and lands are simultaneously turned into waves that modulate it. This type of FM audio technology was used in many of the early computer sound cards from just a few years ago.

As mentioned earlier in this chapter, although the laserdisc video is provided in an analog format, it produces a much greater resolution than VHS videocassettes or typical NTSC broadcasting. This is because the laserdisc features channels that have a greater width than the aforementioned formats. The other formats didn't have enough space to include the extra bandwidth for an improved signal; there wasn't enough TV spectrum space available for broadcast TV and there wasn't enough room on the tape for the VHS format.

Physically, the laserdisc is very similar to a CD. It is a very thin and shiny stamped sheet of aluminum contained within a clear plastic disc. One difference that you can't see is that the audio CD contains two digital audio tracks and the laserdisc has four audio and video tracks. All you can see is a mass of swirls around both discs. Audio CDs look so nice that they would probably be treasured by nomadic tribes, just for their looks (maybe as big earrings or a Star Trek-style necklace?). But because of the laserdisc size, they would be a lot less functional as jewelry in the tribal setting than audio CDs. Rather than being the nice, convenient 4.75 inches in diameter, they are a full 12 inches in diameter (Fig. 4-3). So, they essentially look like a big, shiny record. They would take up about as much space as a record, except that they go into big plastic sleeves for safe storage.

The amount of physical space required for laserdiscs is one of the disadvantages of the format. This isn't a major problem, not like owning a collection of 16-mm films, but the size can still be an

Fig. 4-3 A comparison between the size of laserdisc (top) and a DVD or CD (bottom). Most LD players have separate slide-out drawers to accommodate both LDs and CDs. Pioneer

annoyance if you have a lot. Also, for a standard movie, you will have to flip the disc over part way through. The discs contain only 30 or 60 minutes per side (depending on the type), so if the movie is much longer than the standard 90 to 120 minutes, it will have to be contained on several discs. In this case, you would need to flip laserdiscs three times or more during a single movie.

Compatibility is both a positive and negative issue with the format. Although laserdiscs have become somewhat popular with movie buffs, they have rarely made inroads into the movie rental stores. If you can't rent videos or record in the format, then you will probably either own a VCR and a laserdisc player or just stick with the VCR.

This dilemma slows down the growth of laserdisc sales, which in turn prevents stores from selling or renting laserdiscs. The positive side of compatibility is that some laserdisc decks can also play CDs. So, if you haven't already purchased a CD player, you could just pick up a laserdisc player to handle both responsibilities (Fig. 4-4). Unfortunately, most people who would consider buying a laserdisc player already own a CD player.

Fig. 4-4 This Pioneer laserdisc player, like most, is also capable of playing CDs. The small drawer pops out of the center so that you can insert a CD, or the large drawer can slide out if you feel like watching an LD. Pioneer

The laserdisc format presents plenty of great advantages in video quality, convenience, and permanence. But there are so many gaps in performance, cost, and availability to be overcome between the favorite VHS format and the laserdisc. Now, the circumstances are perfect for another new format to fill the hole and take control of the market. It appears that the answer is DVD.

THE DVD

Right now, it appears that the video technology of the future is the *DVD* (*Digital Video Disc*, although some people are using the absurd name *Digital Versatile Disc*). Unlike some of the previous recent digital audio formats (MiniDisc and DCC tape) and the video laserdisc, the DVD is in position to grab control of a large portion of the video market, possibly even dislodging the VHS VCR format from prominence in the next 5 to 10 years. See Figs. 4-5 and 4-6 for a few different views of a DVD player.

Fig. 4-5 A few different views of a JVC DVD player and a DVD disc. JVC

The DVD is a sort of combination between several different types of digital disc technologies. The DVD isn't just a mutation between formats; it takes the best from everything and produces a sort of all-powerful format that some people call a "super CD." And the names are well-deserved. The numbers are just astounding. The DVD can handle one typical movie—and much more—on the single side of a disc. DVD can also be used for just audio information, handling more than 488 minutes of better-than-CD audio! Unlike the 12-inch laserdisc, all of this information can fit onto one side of a single 4.75-inch disc (the same size as an audio CD).

The secret of the "miracle" DVD is the MPEG-II information compression format. This system was set up by the Motion Picture Experts Group, an international consortium of video industry experts who were called upon to develop a new video standard. The MPEG group was created to prevent a less-than-adequate standard from being passed (as was the case when NTSC was approved) and from "letting the market decide" and having conflicting formats

Fig. 4-6 Another DVD player and its accompanying remote control. Pioneer

battling for years at the expense of the general public (as occurred with Beta/VHS and DCC/MiniDisc).

The MPEG committee first created the MPEG-I standard for noninterlaced video for either CD-ROM or broadcasting applications. One problem with the standard is that the NTSC standard is interlaced, and the broadcast industry is generally opposed to noninterlaced video. The MPEG-II video standard is based on the original, but the second is 20% more efficient. Also, it uses the interlacing technique so that it can work with a system that is somewhat based on current television systems. Another MPEG system, MPEG-III, was developed, but it was scrapped in order to get one standard approved, rather than continue developing standards without ever putting anything into effect.

The key to MPEG standards is compression. Think about it--an entire movie, even "Gone With the Wind," is able to fit on one side of a CD-sized disc. Just 15 years ago, everyone was amazed that 74 minutes of stereo audio could be squeezed onto such a small disc. Now a movie can fit on less than 25% of the area required for it to fit even on a laserdisc. With audio CDs and video laserdiscs, every bit of the picture or audio is represented on the disc. Likewise, every dot on the screen requires a code. This is not the case with the DVD system, where much of the coding is not bit by bit. Instead, much of the coding is based on a change in data, rather than what each bit of data is. For example, if a movie has a blue sky in the background consistently through different screens, a code for a particular shade of blue will be inserted, and a new code will not be inserted until that particular dot changes color. With these MPEG compression techniques, a remarkable amount of information can be accurately squeezed onto a disc.

INTERACTIVE VIDEO

The DVD already has a number of advantages over other video formats, but the biggest changes in store for viewers is the interactivity of the DVD. On any other format, you pop in the movie and watch it. Aside from that, you can stop, fast forward, rewind, or pause it. For a real kick, you can try hitting the freeze frame or slow motion buttons.

DVD includes a huge amount of interactive features. Most DVD videos aren't really even movies; they're computerized movie packages, and the movie is the prime feature. When you pop in the movie, the DVD movie's opening screen pops up and you have a choice of options that you can scroll through. The DVD opening screen is very similar to a menu on a computer Internet Web site. Just click down the list of options and choose one. Some of the choices include language; subtitling; viewing the movie; viewing information about the actors, actresses, and other background personnel; and even special features (such as video segments about how different parts of the movie were created).

These features are astounding. You can choose one of eight different languages for the audio track (all eight different audio tracks are contained on the disc), and you can always use it to practice French for your next trip to Quebec. Also, the movie can be subtitled in any one of 30 different languages (those are also all contained in different parts of the disc). Of course, aside from actually viewing the movie itself, one of the favorite choices for movie buffs is the special features section, where some DVDs include extra chunks of footage from the movie or show how it was produced, etc.

The potential for DVD movies and other video productions is

astounding. It would be possible to produce movies with different endings. The different versions could either be played at random or you could choose which ending you wanted to see. Another possibility is interactive movies or productions, whereby you could enter the data or possibly even make decisions via voice-recognition software. These DVDs could be somewhat similar to some of the computer software currently on the market, except that the video would be clearer than present NTSC standards. These types of discs could provide some incredible educational opportunities. DVD offers still another opportunity for the music video industry. Because DVD is effective as a music-only format, music-video DVDs are being produced. If the format gains momentum, audio-only DVD car head units and portable players could be produced. Then, you could purchase a music DVD for maybe $19, listen to it at work or in your car, and yet still watch the same disc on your home DVD system.

Another crossover benefit for the DVD user involves computers. My mom is astounded because the CD-ROM drive in my computer can play audio CDs. Imagine if she saw a computer DVD drive. One of these units can run CD-ROMs, play audio CDs, and play movie DVDs. It can do almost anything, except record, although recordable DVDs are planned for the near future. This is the ultimate in *convergence*—the meeting of computer, video, and audio technologies. For more information on computer video systems, see Chapter 8.

DIVX: PAY DVD

DVD is a hot topic right now, one that has barely been touched by the mass media, and most people that I've talked with (except for those interested in technology) know little or nothing about DVD at this writing. Yet, not only is DVD moving forward, but it is mutating into other forms.

The most recent form of DVD, which will probably become the most controversial technology since the Internet and the V-Chip is DIVX. DIVX is simply DVD.The discs look the same, hold the sameinformation, and use the same compression techniques. The only difference lies in the accessibility of that information. You can't just pop a DIVX disc into your disc and play it; you must first pay for the number of times that you want to view it. Then, the disc can be digitally "unlocked" and it will allow you to watch it x number of times. You may keep the disc for as long as you want, but any time that you want to watch it, you must pay to have it unlocked.

The DIVX technology is fantastic for video rental stores. If it catches on, they can switch over to DIVX from VHS videotape or laserdiscs. Then, they can just stock a huge pile of DIVX discs. The discs are extremely inexpensive to produce in quantity, so cost won't be a factor. Best of all, they won't have to protect their investment in videotapes or laserdiscs. After someone is finished with a DIVX disc, they can either throw it away or keep it in case they want to have it unlocked to watch again sometime.

The advantages for the video buff are also clear, but so are the problems. First of all, it would be possible for the manufacturers to drop DVD altogether and to go with DIVX instead. Then, you couldn't really own any video--you would have to pay for each play. However, the consumers have some control over the marketplace and I'm sure that DIVX would quickly be rejected if the manufacturers went for such a "power play." (And there's no doubt that blank VHS tape sales would increase and that people would learn to hack DIVX access codes!)

The second problem relates to buying a DIVX disc rather than a DVD disc (or a video of another format). If you would purchase the DIVX, although you would have the disc itself in your posession, you would have no control over your cost to use it. So, if DIVX suddenly jumped from $2 per viewing to $8 per viewing, you would be stuck with it.

With these problems in mind, I believe that the DIVX format has a definite chance to break into the video rental marketplace, but I doubt that it will displace sales from the more "permanent" sources of video information. Still, at this point, we need to determine whether DVD will even take hold, let alone whether DIVX will dominate the market in the years to come.

CHOOSING A DIGITAL DISC SYSTEM

Presently, the choice of digital discs for video is limited to DVDs and laserdiscs. In terms of audio and video quality, both formats should be about the same. After this point, however, everything starts to separate.

The big advantage to laserdiscs is that the format is just beginning to mature. It's been on the market for a few years, and although it isn't common, laserdiscs are popping up in video-of-the-month clubs, etc. DVDs are totally new and nobody really knows just how popular they will become. The big problem with the playback-only format is that if the format collapses prematurely, no other titles will be produced in that format. You will be stuck with a machine that is only capable of playing back titles from a certain small time frame. At this writing, a fairly substantial number of titles are available on laserdisc; even a number of older movies have been released (although the figure is substantially lower than the total number of titles that are available on videocassette). On the other hand, few DVD titles have been released, but they are coming out quickly. At the pace DVD videos are being released, they should catch up with the total number of laserdiscs in the next year or two. But are you willing to take the chance on DVD or hope that the laserdisc will continue rolling, using its place in the market to hold viewers' interest?

As you know from reading the rest of this chapter, I believe that the DVD format has so many positive aspects that it will not only survive in the market, it will become very successful. In fact, I even believe that the DVD will drive the laserdisc right off the market within a few years. At this point, I think that buying a laserdisc player would be unwise. If the DVD can drive the laserdisc off the market, plenty of people will be selling their laserdisc players and videos for a fraction of their list price. Rather than pay the higher price of a new DVD player and software, you might want to go for a used laserdisc player. Or if you already have one, you might feel comfortable knowing that you should be able to find bargain movies for years to come.

Unlike nearly every other technology that's been introduced in the past 30 years, the DVD is starting at a relatively low price. Instead of the $2000 VCR or the $1000 DSS receiver, DVD machines and software are hitting the market at very reasonable (by comparison) prices. The first newspaper advertisement that I have seen offered one DVD player for less than $500 (the same ad also featured DVD movies for less than $20), which is truly remarkable—especially considering everything that the DVD is capable of delivering. With all that the DVD promises and the low initial cost, it shouldn't be long before many people start buying into the format. As this effect snowballs, the prices should drop, causing more people to buy DVD equipment.

Another boost to both formats is the spread of information. Whenever I was researching the DCC audio format, for example, I couldn't find it in any stores. But now, I can hop onto the Internet and find a number of stores that sell both laserdiscs and DVDs. In fact, some of the Web-front vendors even specialize in either or both of the formats.

For example, see:

The DVD Store, which sells a number of DVD players and computer DVD drives:
http://www.dvd1.com

DVD City, which sells a number of DVD players and computer DVD drives:
http://www.dvdcity.com

DVD Digital Domain, a news site that covers all of the latest updates, news articles, and features in the world of digital technology:
http://www.dvddigital.com

DVD Empire, which sells hundreds of different DVD titles:
http://www.dvdempire.com

Aside from the Web space dedicated to commercial vendors, another boost in DVD information on the Internet is from the newsgroups. One source of information in constant use is alt.video.dvd. As you can see, this group is dedicated to the topic of DVD and there's plenty of good information here. Unfortunately, there's also a lot of time-wasting, repetitive, or otherwise useless information here, too. Some of the more helpful posts deal with errors in particular DVDs, new software upgrades and information for DVD players, and personal DVDs for sale. The time-wasters often deal with fights between DVD and LD owners about which format will eventually win the market, which movies are or aren't cool, or with information that is simply incorrect or misleading. If you consult with alt.video.dvd, be sure to verify the advice with others or you will quickly be misled (Fig. 4-7).

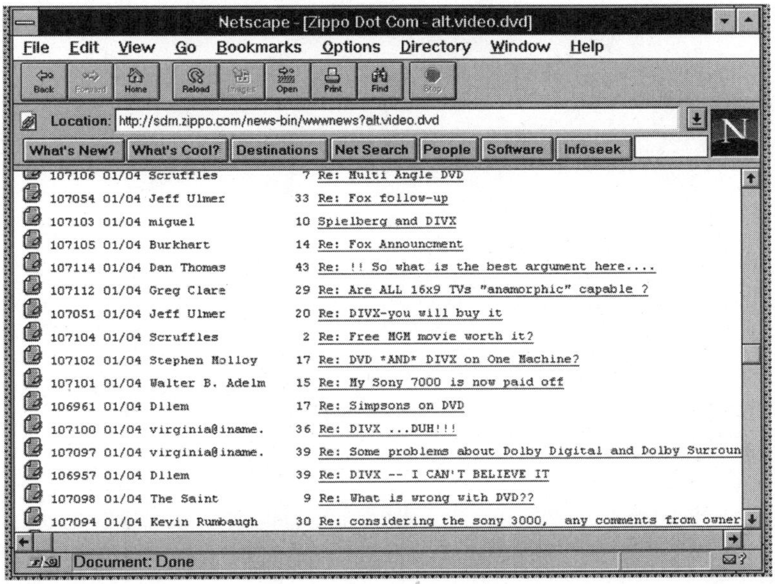

Fig. 4-7 A typical screen display from the newsgroup, alt.video.dvd, a good source of information about digital video formats for the discerning videophile.

CONCLUSION

Although digital discs for video are now just beginning to edge into the mainstream, the technology is maturing to the point where they could become staples of the home theater. Now that you have your video system lined up, it's time to take a look at audio systems in Chapter 5.

5 AUDIO SYSTEMS

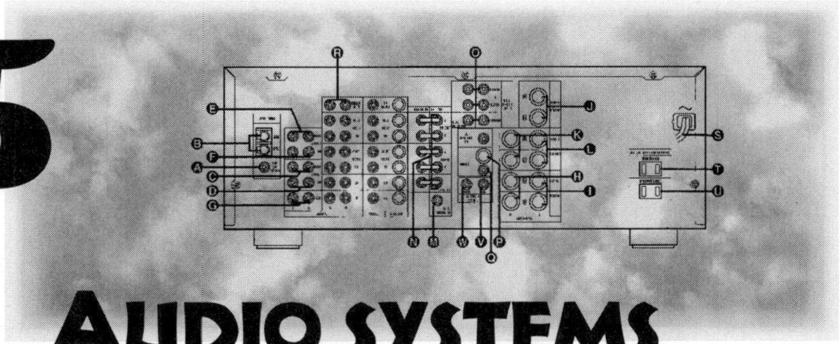

Sometime while you're watching a movie, notice the audio. The voices are crisp and clean, and almost every breath is audible. Background music is used to develop the mood throughout the program. And sound effects saturate the video—often at louder-than-normal levels. Very often, all of the movie voice tracks are re-recorded in a recording studio so that the dialog will sound just perfect.

If the movie studios spend so much time and money to make their audio tracks sound just right, they must be extremely important to the movie's overall effect. Sound adds tremendously to the realism of any video, and no home theater guide could be complete without at least a cursory look at audio systems.

SPEAKERS

Although some elements of home audio and its components might be foreign to you, speakers should be familiar turf. Probably over half the pieces of electronic equipment on the market are equipped with a speaker of some sort. As common as speakers are, there is still a huge difference between the various grades of quality that are available. Some speakers cost as little as 99¢, yet others cost into the hundreds of dollars. What is the difference? Before throwing out some numbers (but not too many), here are a few basics behind speaker theory.

SOME SIMPLIFIED SOUND BASICS

A speaker is just one part of a broad category of components that are known as transducers. Transducers change one form of energy into another. One of the other very common types of audio transducers is the microphone. Understanding what a transducer does is a very important part of understanding how high-fidelity audio systems work and can be improved upon.

Whenever a microphone picks up a sound, it changes sound vibrations into electrical impulses (Fig. 5-1). The sound waves strike a thin element that is called a diaphragm, ribbon, etc. There are a number of methods by which the sound waves are converted, but in one of the common techniques, the diaphragm is held within a magnetic field. Whenever the diaphragm vibrates, it moves within the magnetic field. These movements create an electromagnetic signal that will travel through the line cord. In some cases, the microphones contain preamplifiers to boost the signals up to a more usable level.

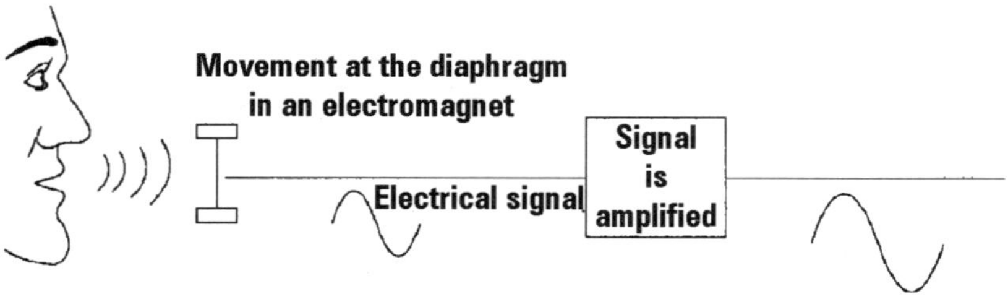

Fig. 5-1 The traditional analog process by which sound is converted into electrical impulses.

With the phonographic record (which is used here because it illustrates the point more clearly than the other methods), the same vibrational pattern that the diaphragm made is translated back from an electromagnetic pulse into physical vibrations. These vibrations are cut into a wax disc, and they exactly match the vibrations that the diaphragm made (with the time factor added). After the records have been pressed, the sound process takes an opposite route. The record needle bounces up and down within the grooves. The depth of the groove controls the volume of the recording and the width controls the frequencies (wide grooves for low frequencies and narrower for higher frequencies). Like a microphone, the phono cartridge contains an element that vibrates within an electromagnetic field when the needle bounces along in the grooves.

These electrical impulses travel to an amplifier, where they are considerably boosted in strength. These signals travel through the speaker wire to the speaker inputs. There, a coil of wire that is known as the voice coil is suspended between a powerful permanent magnet and an iron core. The voice coil is directly connected to the speaker cone (often made from paper), which is what most people assume the actual speaker to be. The electrical impulses cause the voice coil to pump in and out within the magnet and the iron core. Because the voice coil is connected to the speaker cone, the cone vibrates in and out from the center of the speaker, sort of like the action that a piston makes.

As peculiar as it might seem, the audio from a radio, stereo, television, etc., is entirely produced because the thin paper cones are being pushed in and out. Even after years of using speakers and understanding the basic concepts behind how they work, it still amazes me that the thin pulsating paper cone can sound as clear and loud it does. These vibrations from the speaker cones are, in turn, transmitted through the air.

Sound waves, like any form of energy, will dissipate slowly. Fortunately, because of the enormous wavelengths across the voice frequency range, sound has a very limited range. If it didn't, you would constantly be hearing noise from all around the world. Instead, most sounds are restricted to a range of a few hundred feet or less, and only extremely loud sounds (such as explosions) travel for longer than one mile. Chances are that the room in which you will be installing a home theater system will be of an "average" size: anywhere from 10' X 10' up to about 15' X 20'. In these cases, you shouldn't require a more-powerful amplifier and higher-power (rated) speakers. If you have a contemporary home with a large, open floor plan, then you might consider purchasing a higher-power system so that you can adequately cover the house if you want to.

SPEAKER ACTION AND COMPONENTS

Any speaker consists of a number of components that must work together as a team (Fig. 5-2).

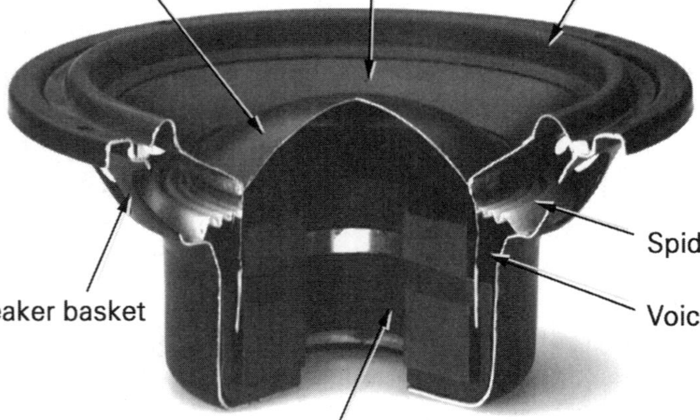

Fig. 5-2 A cutaway view of a speaker with its individual parts are exposed. Morel Acoustics

MAGNETS

As stated in the previous section, the voice coil is a coil of wire that moves within the speaker magnet and an iron core in the center. Because the magnet (Fig. 5-2) is the key to the magnetic field that the voice coil moves in, a stronger magnetic field is necessary for the speaker to be able to handle greater amounts of power. Generally, with the speakers from one particular manufacturer (or from a particular product line), you can go by the rule that the larger and heavier that the magnets are, the better the speakers are. This isn't always the case though, because some magnets have a stronger pull per pound than others. Most speaker magnets are made from ferrite (a form of powdered iron) and ceramic, although some of the speakers that can handle high power have strontium magnets.

So, you can deduce quite a bit from the magnet size and composition, but not everything. That is part of the difficulty with choosing speakers; you can choose by price, by manufacturer, by advice, or by the features in the catalog descriptions, but you will never know what any given speaker sounds like until you listen to it in your house.

VOICE COILS

The speaker voice coil (Fig. 5-2) generally consists of a very fine gauge of enameled copper wire wound around a bobbin, which fits in between the permanent magnet and the iron core in the center (Fig. 5-2). The bobbin must slide straight back and forth within this space without grinding against the magnet and the iron core or it will hang up and fail.

The most common cause of speaker damage results from voice coil failure. When a speaker is driven by more power than it can safely handle, the voice coil will overheat and finally short together. Some high-powered speakers use air-ventilated or liquid-cooling methods to prevent the voice coil from becoming too hot and opening. This is an excellent method to prevent speaker damage. Tweeters are the most fragile of all speakers. The voice coils in woofers and subwoofers must move across large distances, and they are often pumped with huge amounts of power. If possible, purchase tweeters, coaxial speakers, woofers, and subwoofers with liquid-cooled voice coils if you plan to install a high-power system. But the liquid-cooled speakers are more expensive than speakers without additional cooling, so they might be out of your budget.

SPEAKER CONES

Although the voice coils do most of the actual work in a speaker, the speaker cones (Fig. 5-2) account for the sound that you hear.

CONE MATERIALS

Any particular cone material is chosen over another type for reasons of cost, speaker type, sound quality, sound environment, and durability. For example, paper speaker cones are commonly used in home audio systems. However, if speakers with paper cones are used in car stereo applications, they could easily warp from dampness or degrade in the sun's ultraviolet light. Instead, many of the speakers for the car use polypropylene or other resistant material (Fig. 5-3) because they will not warp as a result of dampness in the car. On

Fig. 5-3 This speaker cone consists of woven fiberglass. Polydax

the downside, polypropylene is not touted at the world's best-sounding speaker material, but it lasts longer. So, although speakers with polypropylene cones might be fantastic in your car, you wouldn't choose them for your home.

The materials for speaker cones are also chosen for the frequency range of the speaker that they will be used in. Heavier cone materials are used in the low-frequency speakers and thinner, stiffer cone materials are used in the high-frequency speakers. For example, thick, polymerized paper cones are often used in woofers and subwoofers because the thick material is strong and flexible. Its representation of high-frequency signals is poor, but that's fine because it is only intended to radiate the low end of the audio spectrum. Titanium and other very stiff materials are used in tweeters because they are strong and reproduce shrill sounds very well. These materials cannot tolerate the low-frequency sounds because they would be quickly destroyed by the long, deep movements that are required to reproduce sound at these wavelengths. Because of the medium range of the midrange drivers, they are often a compromise in materials between the woofer and the tweeter.

You can't effectively choose your speakers based on materials alone. So many different factors are involved in the final sound (everything from the magnet, enclosure type, and cone materials to the engineering design of the system) that the only way to choose a speaker system is to (once again) just listen to it.

DUST CAPS

Dust caps are merely dome-shaped caps that are sealed around the center of the speaker cone over the area of the voice coil (Fig. 5-5). The dust cap is primarily intended to prevent dust, dirt, and debris from entering the small space around the voice coil, and consequently, destroying it.

SURROUNDS

As the speaker cones are pumping like pistons in a car engine, the surround is flexing and holding the cone to the speaker's frame (Fig. 5-2). The *surround* consists of a thin ring of polyester foam, polyether foam, butyl rubber, paper, or cloth that attaches the outer edge of the speaker cone to the top outer edge of the metal basket. The surround helps to control the speaker movement and it also absorbs energy from the speaker cone. This combination is flexible enough to allow the speaker cone to move and produce high-quality sound, yet it is tough enough to keep the cone in place without tearing away.

SPIDER

Contrary to unpopular beliefs, the speaker *spider* was neither named after a Fiat or because this form of suspension is made out of ground spiders. Frankly, I have no idea why this ribbed, circular, resinous piece of cloth mesh is so named (Fig. 5-2). The spider of a speaker is a piece of cloth that has been soaked in resin, so it holds its shape, yet is still flexible. Like the surround, the spider is connected to the speaker cone and the speaker basket. Likewise, the spider also helps to control the speaker cone and prevent it from distorting its shape (and, in turn, its sound quality).

SPEAKER BASKET

The *speaker basket* looks and functions like the name implies—it is a metal basket that holds the speaker components (Fig. 5-2). If the speaker was an animal, the basket would be the exoskeleton; it provides the necessary strength and rigidity to hold the entire unit together (and it is all on the exterior). It is important for the speaker baskets to be constructed out of one piece of metal to prevent any rattling at high volumes.

SPEAKER CONNECTIONS

Most speakers have a pair of lugs that are mounted on a small piece of circuit board material. This board is then mounted on the side of the speaker basket. Nearly all speaker companies use standard lugs that mate with slide-on connectors.

SPEAKER CLASSIFICATIONS (BY FREQUENCY)

This section covers the actual speakers that are used inside an enclosure. It is important to know the function of speakers if you are building your own speaker enclosures and need to purchase your speakers separately. Or you might be nosy and want to know exactly what's behind the grille cloth of that great-sounding walnut cabinet.

To some extent, speakers (themselves) can be classified by size or maybe even by the materials that they are built with. But, a more accurate method of classification is to sort the speakers out by the frequencies that each model is most capable of reproducing accurately.

Other than woofers and subwoofers, no speakers can fit into more than one classification. It just wouldn't be accurate for a speaker to be listed as a "tweeter/midrange/woofer." In that case, it would simply be a midrange.

TWEETER

A *tweeter* is a speaker that is designed to reproduce high-frequency audio (Fig. 5-4). The general frequency-reproducing specifications of tweeters are from 4 to over 20 kHz. One of the most important actions of a tweeter's speaker cone is to move very quickly—otherwise, it will not be capable of reproducing the high

Fig. 5-4 A tweeter with a textile dome. Polydax

frequencies. Because of this need, tweeter cones are often made of stiffer materials than those that are intended for lower frequencies. Tweeters must be used in conjunction with other speakers.

MIDRANGE

Midrange speakers are designed to reproduce audio in the middle frequencies (Fig. 5-5). Midranges are a bit less interesting because the woofers and tweeters operate in a specialized manner, but the midranges are much more of a compromise between the two designs. These frequencies overlap a bit with the tweeters and woofers because every speaker varies a bit in the styling and the frequency range that it is intended for.

MIDBASS

Midbass speakers aren't especially common, but they are designed to reproduce audio in the upper end of the woofer region or lower end of the midrange region. Most midbass speakers cover anywhere from approximately 200 Hz to 3 kHz, which is anywhere from the bass region up through much of the midrange frequencies. Overall, most midbass speakers are designed in the same general style as woofers, except smaller.

Fig. 5-5 A midrange loudspeaker. Polydax

WOOFER

Woofers reproduce bass frequencies, generally in the range from approximately 30 Hz to 2 kHz (however, they do much better in the lower end of their collective range). Woofers are characterized by their large sizes and heavy-duty construction. As was mentioned previously, the speaker cone materials are often (but not always) thicker and tougher than those for high-frequency applications. The overall favorite material for woofers and subwoofers seems to be paper or some combination of paper and another material.

SUBWOOFER

A microcosm of woofers, *subwoofers* reproduce the low bass frequencies, generally in the range from approximately 20 Hz to 1 kHz. In many cases, subwoofers are not considered to be a separate grouping, but are instead considered to be a subset or just lower-range woofers. There is some basis for this lack of a separate grouping because the frequency specifications for subwoofers are not significantly different from those of woofers, and also because the size and construction of the two types are very similar.

The general division between woofers and subwoofers in home audio systems is that woofers are included in an enclosure with the midrange and tweeters. Subwoofers are like your weird Uncle

Fred—they're a part of the speaker family, but they're always stuck off in their own enclosures, by themselves. In many cases, subwoofers are large, and their enclosures require a sizable chunk of space, too. Often, subwoofer boxes are designed as a big, stylish wood box or sometimes as an ottoman—anything to take your eyes off the huge "thing" in your living room.

OTHER SPEAKER SPECIFICATIONS

In addition to the frequency ratings, several other ratings are important when you are considering a group of speakers to purchase for your system. The power specifications can be even more crucial than the frequency ratings, so read and choose carefully.

INPUT SENSITIVITY

The *input sensitivity* is the sound pressure level in dB that is output when 1 W is delivered to the speaker. This measurement is taken at a range of 1 meter from the front of the speaker, at a given frequency (usually 1 kHz). The input sensitivity rating is only important if you are building your own speaker system.

RESPONSE

The *response* of a speaker is a graph of its output across the entire frequency range. To understand speaker response, imagine that the input sensitivity measurements for a speaker were taken across the entire frequency range, then all of those measurements were listed on a graph. That's it. The best speakers will have a plateau-shaped response with a fairly sharp roll-off on either end. Most speakers have somewhat sloping responses, etc., but as long as they are relatively flat, they should do fairly well.

NOMINAL POWER HANDLING

The *nominal power-handling rating* of speakers is the amount of power that a speaker can handle continuously. The nominal power-handling rating is the practical power limit that you should ever consider inputting from your amplifier. For that matter, if the amplifier is driven at high levels, a system is much better off with speakers that are rated approximately 10% higher (in watts) than the power output of the amplifier, just to be safe (e.g., if you have a 100-watt amplifier, your speakers should be rated to handle at least 110 watts).

PEAK POWER HANDLING

Not to be confused with *nominal power handling*, the *peak power handling rating* is the instantaneous amount of power that a speaker can safely tolerate. The peak power-handling rating is more of an academic figure than one that you could use to design a system. In many cases, you will not know the output levels of your amplifier on peaks while you use it, unless you have electronic measuring equipment. Rating a system by the peak power and not the nominal power would be pushing the system for a little more. It's much safer just to use the nominal figure and play it safe.

SPEAKER POLARITY

If you look closely, you will notice that every set of speaker terminals has a + sign on one side and a - sign on the other side. Likewise, the amplifier will also either have these same markings or a red terminal (positive) and a black terminal (negative). These markings are to enable you to match the polarity of the speakers to the amplifier and to the other speakers, if they are connected in series or in parallel.

Speaker polarity is one of those "if it sounds good, do it" subjects. Connect your speakers and if you think that they don't sound quite as good as they should, try reversing the connections and listen for a difference. Chances are that you will notice only a slight difference, at best.

MULTIPLE-DRIVER SPEAKERS

Multiple-driver speakers are typically used in car audio applications, where space and mounting locations are minimal (like trying to buy real estate in Manhattan). Because a house should have plenty of available space (like trying to buy real estate in rural Nebraska), the manufacturers don't bother with combining several speakers into one unit.

Multiple-driver speakers have several speakers that are built into one basket. The most common type of multiple-driver speaker is the three-way arrangement, which has one midrange and two different frequencies of tweeters built onto the frame above the midrange cone. Other possibilities are the standard two-way (coaxial) speaker and even the four-way speaker.

Typically, speakers sound a bit better when they are separated. So, the auto multiple-driver speakers don't sound quite as good (as good as separate speakers), but they are built to be physically and electronically rugged. Although parasitic tweeters are occasionally used in home audio applications, multiple-driver speakers are almost never designed for home use.

In theory, there is no reason to consider multiple-driver speakers for use in your home because their benefits are mostly lost. However, like anything, there are a few exceptions. The major benefits of using old car speakers is size and cost. I know I just said that size wasn't relevant, but hang with me for a minute. Because cars are wrecked from time to time and audioheads constantly upgrade their systems, some good car audio box speakers (don't just buy the speakers unless you plan to also build the enclosures) are often available in classified ads for a nice price—often less than half of the original.

If you are on a tight budget and live in an apartment, a good, used set of car box speakers is the way to go. The sound will be pretty good, the price is great (probably $25 to $50 used, unless you happen to have your own set around), and the speakers are small and easy to move. A better choice would be a set of Baby Advent or bookshelf Bose home audio speakers, but the only way that you will find them in this price range is if you get lucky at a pawn shop or find someone who is bankrupt and needs money FAST (more on buying speakers in the next section).

The second application for multiple-driver speakers in the home is when attempting to provide surround sound on a budget. You might already have a decent component stereo system that cost about $700 to $1200. If you have the money to add a surround-sound kit to your system (typically about $200 to $500), then go for it. But, you might consider using used car box speakers because the price is great and these speakers will typically do a fine job of filling background audio. However, beware of car audio speakers if you have a light-duty amplifier or if you often blast the volume, pushing the amp to its limit. Home audio speakers were designed to operate at 8 ohms of impedance and car speakers run at 4 ohms. The end result is that car speakers will make the amplifier work harder (at the same turn of the volume knob) than 8-ohm speakers.

THE SPEAKER COMBINATION

Speakers that were assembled specifically for home theater applications differ from stereo system speakers. The most common stereo speaker configuration is to use two three-way speaker cabinets. The three-way system features a tweeter, midrange, and a woofer in each cabinet. With stereo systems, the emphasis is on the higher- and lower-frequency audio because many instruments emit quite a bit of sound in these frequency ranges.

On the other hand, home theater systems typically usea pair of front speakers (each containing a midrange and a tweeter) and a pair of smaller rear speakers (each containing a midrange and a tweeter), a center speaker (which contains either one or two midranges), and a "subwoofer" box (which contains either one or two woofer or subwoofers).

The reason for the difference in speaker types between typical stereo systems and home theaters lie both in the content of audio and the way in which the speakers are listened to. The goal of a typical stereo system is to reproduce the sound of being at a concert. There, the music will be performed in front of you, so the main concern is keeping the audio balanced between the left and right channels (Fig. 5-6A).

Fig. 5-6A A typical microcomponent stereo system with two box speakers. Because this is a smaller system, it only includes a tweeter and a small woofer in each speaker cabinet. JVC

In the home theater, the first concern is to make the system surround you (Fig. 5-6B). This is important because sounds from real life should envelope you from all sides, not just the front. Just because of the way that our ears are situated on our heads, it makes sense that the rear speakers could be less powerful than those to the front. A subwoofer is used to improve the thunderous bass for explosions, crashes, and more.

The last element of the home theater speaker system is the center-channel set of midrange speakers(Fig. 5-7). As is mentioned elsewhere in this book, an imaginary center channel exists if the speakers are correctly aligned. In a good listening environment, the center channel is unnecessary and it could even negatively alter the sound stage. I suppose that the manufacturers assume (probably correctly so) that most home video equipment will be moved into a pre-existing living room environment (rather than "build" the living room around the audio/video). If this is the case, they just automatically include the center channel to be sure that everyone has a stable sound stage (even if the stereo imaging is limited).

Fig. 5-6B A small set of home theater speakers: subwoofer (center, bottom), center channel (center, top), main speakers (on either side).

BUYING SPEAKERS

Buying speakers can be a big problem because, as far as sound quality is concerned, speakers are the most important components in your system. Yet, you can't necessarily rate speaker quality by looks, size, size of the magnet, speaker cone materials, or price. Because of the variables in speaker construction and design, it is somewhat common for most people to give a better rating to a speaker that costs twice as much as another. Notice that I said "most." Beyond a point, good-quality audio is completely subjective. Although the ultimate goal of most people is to have a system that exactly reproduces the music that was produced as it was being

Fig. 5-7 A typical center-channel speaker cabinet for home theater applications.

recorded, some actually would rather have a system that is inaccurate. One example of this is the action-movie fanatics who would rather boost the bass higher than the proportion at which it was recorded, so that they can feel the action as well as hear it. When it gets down to intricacies such as these, it's your system and you have to buy it. Pick the sound that you like.

Similar to the situation with amplifiers, one of the worst problems with picking out speakers is the sheer number on the market. Literally hundreds, if not thousands, of different enclosed and separate (raw) speakers are on the market from year to year. Which ones will you pick?

When it comes right down to it, the main factors that will affect your choice of speakers are cost and availability. With so many different speaker varieties available, chances are that you will not be able to find some of the brands and models. No stereo stores have the facilities to buy every model that's on the market; very few would even have 1/20 of those in stock. Pricewise, chances are that you will get the most for your money from the large companies that sell in such volume that they can afford to offer lower prices or to offer special discounts to the distributors who buy many of their units. Of course, many audio enthusiasts will also note that some of the larger companies use their big name to sell substandard equipment. So, where do you turn to?

If you don't really care about audiophile-quality sound, you might just plan to purchase a system from one of the major names in consumer stereo equipment (such as Technics, Sony, Kenwood, etc.) and use the speakers that come along with the system (Fig. 5-8). Taking this route will typically provide you with a decent system for the price. But, chances are that you would have been happier with the final result by assembling the system yourself. If price is a factor, shopping around for closeouts, floor models, and used equipment is definitely a help.

Fig. 5-8 This "do-everything" home audio system is already equipped for home theater applications, with an A/V receiver, a set of rear-channel satellite speakers, and a center-channel speaker. Pioneer

On the other hand, if you are concerned about the sound of your system, research anywhere you can and gather as much information as possible. Read the magazine reviews and tests for all of the speakers that might catch your interest. Unfortunately, many of the audio magazines on the market specialize in expensive, high-end equipment. *Stereophile*, for example, rarely reviews speakers that cost less than $1000 per pair. It can be difficult to even find a pair of speakers for less than $1000 in *Stereophile* classified ads! For more affordable audio systems, check the prices in *Stereo Review*.

Next, go to audio stores and listen to as many different speakers as you can. These shops have their own sound rooms so that you can test out the various speakers on your own. Of course, the speakers will all sound differently when they are mounted in your own house, as opposed to being mounted in the wall of a big show room. Still, they will give you an idea of what you can expect from a set of speakers. Be sure that you go to each testing session adequately prepared. Take a few examples of music that has excellent audio and recording quality and that you know very well. Listen closely to the various parts of the songs and see how well the speakers reproduce the sound. Listen to the tracks at home in tandem to your listening test trips when possible. Unless you have been tweaking audio systems for a few years, you will probably not have the experience to automatically know what a system is lacking.

While you are at the stores, ask the salespeople as many questions as possible. Ask questions about system designing, repairs, warranty information, and anything else that might come to mind. Also ask for opinions because chances are that the salesperson knows these speakers better than you do. Remember that anything that the salesperson says is going to be colored by the fact that he or she will be getting a commission from the sale. Salespeople are biased toward any product that their store carries, and they will generally lean toward the more expensive products that they sell. It is very rare, but if the salesperson recommends equipment that is sold elsewhere, be sure to consider the advice; the salesperson is going against store policy and personal interests to give you honest advice.

If you have a computer, be sure to check into the Internet Usenet groups for some decent audio information. Aside from the Usenet groups, there are also some excellent Web sites on the Intenet. The first sites to visit when researching your next purchase are:

http://home1.swipnet.se/~w-10118/hifi This site is an enormous, comprehensive listing of other audio-related sites on the Internet. The sheer volume of audio sites is amazing!

rec.audio.high-end Get ready to shell out the big bucks! If you want to get some information on those hand-built custom speakers from the tiny village in Denmark, visit this site.

rec.audio.opinion More opinions on audio systems; this group is touted as an opinion group, not a simple newsgroup.

Like any of the Usenet groups on the Internet, expect plenty of arguing. Part of the problem is that audio is a subjective hobby and people have different standards to follow. This alone causes plenty of arguments and disagreements. Also, people generally are not very nice. Some people are just bent on arguing, no matter what the topic. The following advice is common sense: If a subject line contains expletives, don't bother to read the message. It's not worth your time. I wouldn't even mention it, but I've been enticed by some subject lines that read something to the effect of "Brand X speakers are &%$@ed up!!!!!!!!!!!!!!!!!!" I would check it out and the message would be probably either wrong or misleading. Worse yet, you might catch a response to this sort of message and discover that the 47 different messages on the same subject are actually an argument about the possibilities of a communist takeover in the United States or about the musical integrity of the Spice Girls.

The bottom line is that although the Internet is a fantastic wealth of information, in the worst case, it can be a misleading waste of time. When you investigate on the Internet, have a game plan of what you want to learn, be very perceptive, and keep an eye on the clock.

SPEAKER PLACEMENT AND ROOM ACOUSTICS

The placement of your speakers, the size and shape of your listening room, the surface materials used to build the room, and the objects in the room all affect the final sound of your stereo system. For more information on speaker placement and system design and installation, see Chapter 7.

A/V AMPLIFIERS

The speakers might be the voice of the home theater system and the TV might be the eyes, but the A/V amplifier is the heart. The A/V amplifier operates as the control center for the entire system. All of your components (the TV, VCR, and DVD or laserdisc) get plugged into the A/V receiver. Here, you can control everything in the home theater system (Fig. 5-9). Because of all of the functions that are now stuffed into a single A/V receiver, the name is really a misnomer. The A/V receiver is first and foremost a control box for both audio and video functions. The amplifier aspect is just a secondary function; you could get a power amplifier with just an on/off switch and a volume knob to amplify the signals.

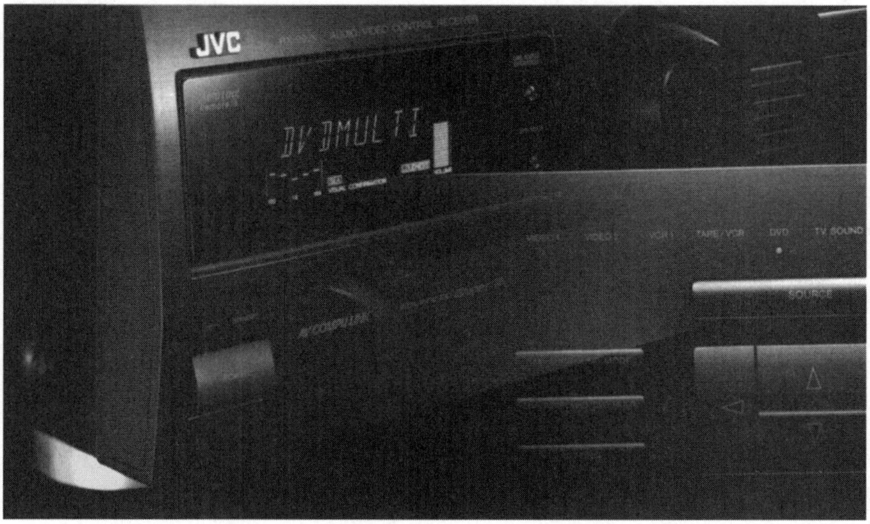

Fig. 5-9 A few different views of an A/V receiver, showing the display and some of the controls . JVC

Another point of confusion concerning the *A/V receiver* name in a home theater system is that it isn't really ever used for receiving

Fig. 5-9 (Continued) Two more A/V receivers. Yamaha (top), Sherwood (bottom)

anything. In a video system, the only receiver is either the television or the satellite receiver. The *receiver* part of an A/V receiver is only used when receiving AM or FM radio stations. But, for some reason, those in the hi-fi world very often refer to receivers as *tuners*--even though this name is not used in conjunction with any other type of receiver. Confusing, eh?

The following sections are primarily dedicated to covering all of the features and most of the ratings that are included in modern A/V receivers. The toughest part of shopping for an A/V amplifier is deciding which features are important to you and which are superfluous. There are so many to consider that making a decision can be tough. See the end of this chapter for some information on making an overall decision. For more information on wire and wiring, see Chapter 6; for more information on system installation, see Chapter 7.

A/V RECEIVER ALTERNATIVES

As mentioned in the previous section, A/V receivers are primarily used as a control box, to switch between the various audio and video sources, and as an amplifier to control the loudness of the signal. But, there are a few hgher-end alternatives to simply going with an A/V receiver.

One of the problems with electronics relates to shielding and interference. Electronics circuits don't always perform up to their typical standards when in close proximity to other circuits. Some of the common end results from this sort of electrical interference are hum (including electrical "buzzing" sounds) and reduced audio signal response because some of the audio signals cancelled each other out.

The solution to the problem is simple, but expensive: Electrically shield the equipment and build it in separate stages. Better yet, keep it simple--adding any gadget increases the chances that the signal will be negatively altered. This advice flies in the face of decades of work in modern consumer electronics. For years, the trend has been to make it smaller, combine many functions into one box, and to add as many gadget circuits as possible.

In home theater applications, this often means separating the A/V receiver into two units: an A/V preamplifier and a power amplifier. This way, the power-amplification circuits and their heavy-duty power supply can be shielded from the sensitive A/V control circuitry. The A/V preamplifier (Fig. 5-10) is essentially what you would think of as being a standard A/V receiver. The only difference is that it doesn't contain the power-amplification. So, just from looking at a unit, the only way to know if it is really a preamplifier is to read the front panel. The amplifier, on the other hand, is usually a heavy box with a power button and a power-indicator LED, although some also include meters or volume controls (Fig. 5-11).

Fig. 5-10 An A/V preamplifier. Rotel

A subdivision of power amplifiers is the mono amplifier. In high-end applications, the mono amplifier is sometimes used as one more step in the separation of equipment. In this case, the left output of the A/V preamplifier would run to one mono amplifier and the right output would run to another mono amplifier. Because of the cost of purchasing two separate high-power mono amplifiers (which can run into the thousands of dollars apiece!) and an A/V preamplifer, this arrangement is not common.

Fig. 5-11 A power amplifier. Rotel

A more-common arrangement is when the mono amplifier is used as a subwoofer amplifier. The subwoofer amplifier (Fig. 5-12) is used both in high-end installations (with an A/V preamplifier and a power amplifier) and simply with an A/V receiver. Even though subwoofers typically only sound like they add a little punch to a system, they require lots of power and can make an A/V receiver

Fig. 5-12 The back of a mono amplifier, which is often used to power subwoofers, showing the inputs and outputs. Rotel

or power amplifier run hot. So, this amplifier is used to take the burden off the system and allow everything to be played louder.

If you're considering a separate subwoofer amplifier and you have a set of home-theater speakers with a subwoofer, check to be sure that an amplifier isn't already built into the subwoofer box (If it has a built-in amplifier, it will have a power cord.). See page 142 for more information concerning mono amplifiers, power output, and amplifying subwoofers.

POWER RATINGS

Power ratings for dedicated power amplifiers are the one piece of technical information that consumers base most of their purchases on. And, it makes sense. Sound quality is subjective. You just want to fill your home with good-quality sound (or perhaps blast music through the neighborhood). Either way, the major concern is the sound level from your speakers. You shouldn't have to worry about the sound quality because the amplifier merely boosts the sound level, right? Yes and no.

Power amplifiers do color the sound that passes through them. Because most amplifiers are designed fairly well, choosing a fair-quality amplifier won't make your system sound bad, but it could make an otherwise great-sounding system merely sound good.

Most people think of amplifier wattages as being linear. When they buy, they get a 100-W amplifier because it is twice as loud as a 50-W amp, and so on. The problem is that sound levels are not based on a linear system of measurement. Instead, sound levels double, based on orders of magnitude. For example, if your amplifier is rated to output 100 W, adding another 100-W amplifier will not double the sound pressure level, which is normally measured in decibels (dB). Instead, to double the loudness of the system, you would need to replace the 100-W amplifier with a 1000-W system of amplifiers!

As you can see by the difference between wattage and actual sound levels, you will not normally notice the difference between a few watts here or there if the systems are rated for over 25 watts per channel. More than likely, you will notice more of a difference in loudness just from the different means that manufacturers use to determine the amplifier power ratings. To inject some advice at this point, it is better to purchase a high-quality medium-power amplifier than it is to buy a cheaply built high-powered amplifier. Chances are that the audio will be a bit better on the better amplifier, it will have more features, and it might be rated for less power than it is actually capable of (although the "high-powered" amp might actually be capable of less power than it is rated for).

Depending on how the equipment is rated, the amplifier could sound differently than you would expect for the level at which it is rated. For example, most audio amplification stages will distort at an increasingly higher level if they are driven beyond a certain point. For the 2 X100-watt amplifier that you just picked up, do you know what the percentage of distortion was when those 100 watts per channel were measured? Perhaps it was 0.01% or perhaps it was 10%. The difference is incredible. If the distortion was only 0.01%, it should still sound good above 100 watts per channel. But if the distortion was 10% at 100 watts out, the actual usable power might be more like 50 watts per channel, but considering the level, it would probably be useless at any output power. That amount of distortion is just too great, even for listening by people who aren't interested in good audio.

Thus, the rated amount of power in an amplifier depends, in part, on how much distortion the designers believe is allowable. Not only is the maximum amount of allowable signal distortion subjective, but it is not based on a standard amount. As a result, some amplifiers perform as the examples in the previous paragraph: one 2 X100-watt amplifier might be capable of well over 100 watts per channel, although another 2 X 100-watt amplifier might actually only be capable of 50 "clean" watts per channel. The first time I noted this difference was when I went shopping for a home stereo tuner/amplifier when I was in high school. I had to choose from some major brands with many extra features and basic offerings from a higher-end brand. The salesman showed me how the 50-W per channel amplifier actually didn't sound as loud as the conservatively rated 35-watt-per-channel amplifier, in part because the former was much noisier and it began to distort after a certain volume. Incidentally, this story isn't just based on good salesmanship; the amplifier that he was pushing was a cheaper model than the one that he was telling me to avoid.

SHOPPING FOR AMPLIFIER SPECIFICATIONS

So, how do you know what to expect from an amplifier? The last thing you want to do is search for an amplifier that fits your needs for power, only to find that it can't deliver the power that you purchased it for. As you might suspect, these power ratings can be used to deceive consumers, and unfortunately, some companies seem to purposely mislead their customers. Very few people have the resources to be able to test and rate the amplifiers on the market for power ratings and sound quality.

Average output power is only one way in which amplifier power can be rated. The ratings covered so far have generally been for average output power. Amplifiers, as well as most other electronic devices, can also be rated in average input power, peak input power, peak output power, and several other means. Input power is the amount of power that is consumed by an amplifier. Depending on the class of amplifier that is used, the amount of output will vary considerably from the amount of power that is input to an amplifier. As a result, input power is a nebulous term. Input power is not frequently used to describe audio amplifiers, but it is a common manufacturers' specification for amateur radio transmitters and amplifiers. Breaking this term down further, peak input power is the maximum input power that the amplifier can safely handle, and average input power is the average amount of power that is used by an amplifier during its course of operation. Likewise, average output power is the average amount of power that an amplifier outputs to the speakers during its course of operation, and the peak output power is the maximum power that the amplifier can safely deliver to the speakers. Because the only rating that matters is one based on what you hear, only average output power is useful for determining how much power you need for your system. Peak and average input power are both important for determining whether your electrical system can tolerate the power-eating requirements of your amplifier. Peak output power can be used to determine what power ratings you must have for your speakers so that they won't blow out on a peak.

NOISE

One problem that occurs during the process of amplification is noise. This term can easily be confused with distortion, which occurs because the amplifier alters the signal as it amplifies it. Noise is sometimes called *line noise* because it is a low-level sound, usually a hiss or maybe a hum, that is present in the source and is amplified by the amplifier. Also, some line noise develops in the amplifier because no electronic circuits are perfect, although digital audio certainly improves upon this problem. You can check various amplifiers in this regard. Just find the rated specifications for the *signal-to-noise (S/N) ratio* of the particular amp.

You should look for the highest S/N ratio possible because that is the ratio of actual amplified signal to the amount of noise that is developed in the amplifier. The signal-to-noise ratio is rated in decibels (dB), so it is quite easy to determine which amplifier has the best ratio—just look for the highest numbers!

The only problem with this rating is that the rating must have a specific reference or the figure is worthless. The typical reference signal for amplifiers is one watt. If any figure above one watt is used, the number of dB above one watt must be subtracted from the listed S/N figure to find the standard S/N ratio. If you are worried about signal-to-noise ratios, check the equipment box for manufacturer's information, reviews in the audio magazines, or call the manufacturer.

DISTORTION

One of the figures that you will commonly see listed in the amplifier specifications is *THD (total harmonic distortion)*. Any amplifier will cause some distortion to the audio signal. In the case of THD, the amplifier distorts the audio by producing harmonics of the input signal. These harmonics are not "clean" representations of the audio; rather, they make it sound more ragged.

Like the signal-to-noise ratio, the THD depends upon a reference (in this case, a reference power and a reference bandwidth) or the figure is virtually useless. The standard range that THD is measured in stretches across the entire audio frequency from 20 Hz to 20 kHz. Most amplifiers will have a higher or lower THD at different frequencies, so if the THD is only rated for one particular frequency or for one portion of the entire audio frequency range, beware. The THD will most certainly be higher for other frequencies inside the audio range because any manufacturer that lists in this manner would pick a low THD in order to get a better figure.

MONO SUBWOOFERS

The wavelength of low frequencies is longer than the entire length of a typical room (e.g. the wavelength of a sound wave at 50 Hz is 22' or 6.7 m). As a result, the sound pressure wave is virtually the same at all distances, compared to the wavelength (22' or 6.7 m, as was given in the previous example). For that reason, stereo subwoofers in most rooms are virtually useless. The benefit of the "directionless" bass is that you can install a subwoofer most anywhere in the room and not have to worry about the sound noticeably emanating from one particular location. Considering the size of most subwoofers and their enclosures, this is a great relief!

For lower powered systems, you can often bridge a subwoofer across the negative side of one of the stereo output channels to the positive side of the other output channel. Check your amplifier or the one that you are considering purchasing to see if it is possible to bridge it in this manner. If you can't do it, then a dedicated subwoofer amplifier might be in order. Even so, the high-powered systems require separate subwoofer amplifiers because bridging the subwoofers across the outputs of the other amplifiers probably won't drive the subwoofers properly. Another downside to bridging subwoofers across the outputs of the stereo amplifier is that you lose the volume control on the bass. With a dedicated woofer or subwoofer amplifier, you can tweak the low frequencies to the levels that sound great to you.

What this all means to you is that you might consider whether the amplifier has a subwoofer output. Otherwise, you might want to purchase a separate amplifier just to power the subwoofer (or purchase a self-powered subwoofer). In most cases, separate mono amplifiers are extremely expensive because they are intended for audiophiles and are built to very high specifications. Some inexpensive kit mono audio amplifiers are available for the electronics hobbyist. In most cases, however, the powered subwoofer is probably the way to go, unless you enjoy playing with electronics. These units are much easier to work with and much less expensive than purchasing nearly any given mono amplifier.

FEATURES

Although A/V receivers and the names of different features vary from model to model, the following list of features covers much of what is available. It should give you a good idea of what to look for.

volume The volume controls the amplitude (how loud or quiet the sound is) of the signals being output from your amplifier. The volume control is a potentiometer (variable resistor) that is typically a rotary unit (knob), although some amplifiers use slider controls or pushbuttons (on remote-control A/V receivers).

bass The bass control is a simple equalization knob or switch. It can boost or weaken the bass frequencies that are being amplified by the amplifier.

treble The treble control is a simple equalization knob or switch. It can boost or weaken the treble frequencies that are being amplified by the amplifier.

balance The balance control is a knob or slider control that allows you to shift the sound to different speakers (left or right). Typically, the balance is a linear control, so that you can shift the appropriate amount of sound to the left speakers and right speakers. In a perfect system, a balance knob would be unnecessary, but it is required because speakers cannot always be placed so that each will perfectly output sound to where you are sitting.

loud function The loud (or loudness) function is fairly typical on home and mobile audio systems. The loud button functions as a bass and midrange booster; press it and the music punches through with

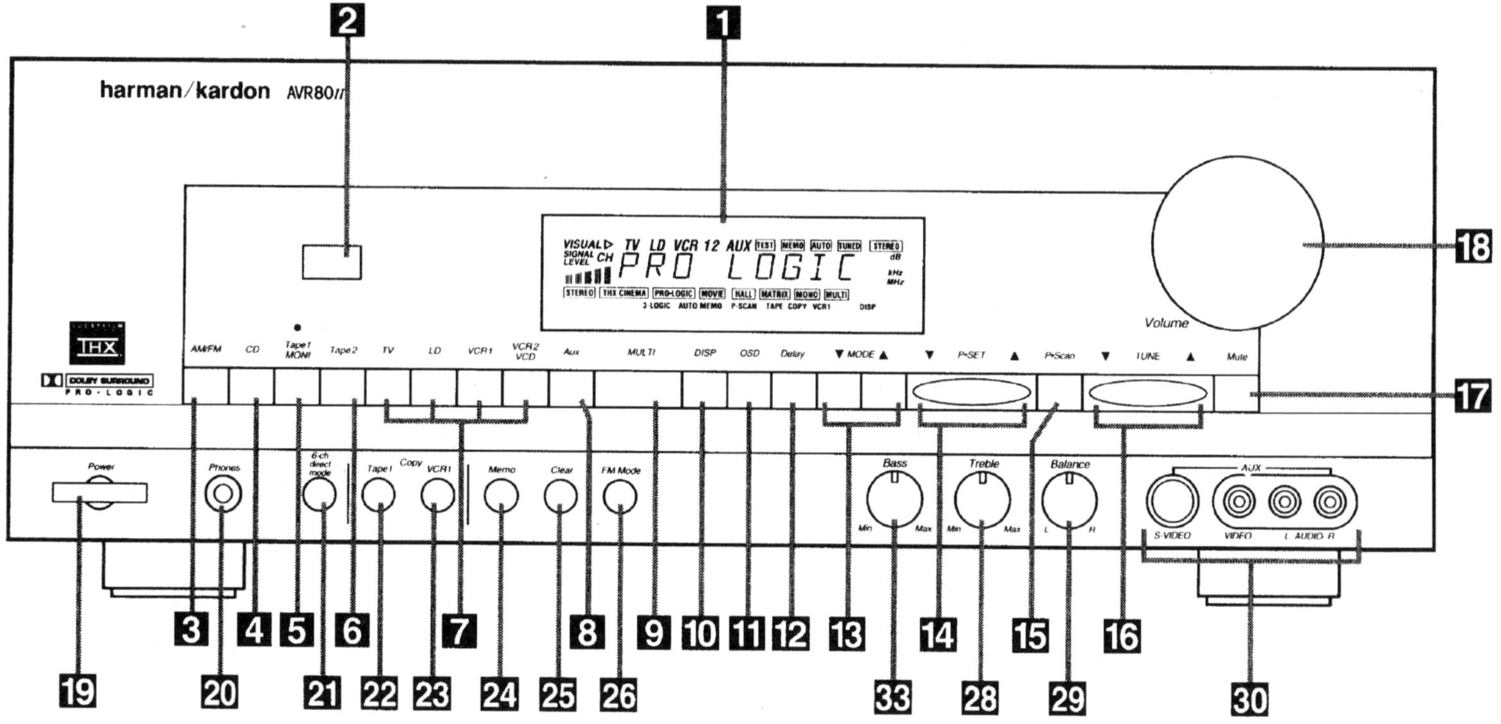

Fig. 5-13 The front panel of a Harmon/Kardon A/V recever, showing its many features: 1. Information delay, 2. Remote sensor window, 3. AM/FM tuner mode selector, 4. CD, 5. Tape 1 monitor, 6. Tape 2, 7. Video sources, 8. Auxillary, 9. Multiroom audio select, 10. Display, 11. On-screen display, 12. Delay, 13. Mode, 14. Preset tuning, 15. P-scan, 16. Tune, 17. Mute, 18. Volume control, 19. Power, 20. Headphone jack, 21. Six-channel direct, 22. Tape 1 copy, 23. VCR 1 copy, 24. Memo, 25. Clear, 26. FM mode, 27. Bass, 28. Treble, 29. Balance, 30. Front-panel inputs. Harman/Kardon

a much thicker sound. Human hearing has a much higher threshold for audibility at the lower frequencies. As a result, if you normally listen to music at low volumes, it will sound thinner than if you rocked the neighborhood. To compensate, manufacturers added this circuit so that music could be faithfully reproduced, whether or not the volume was high or low. Because of its purpose, the loud function should not be used if the volume is already cranked up.

stereo/mono The stereo/mono switch controls whether the system output is the same or different between the left and right channels. The major typical application for this function is when listening to radio stations. If the signal is fairly strong, but not solid, it will sound very hissy, occasionally breaking up. To remedy the situation, just switch over to mono and the signal will be much clearer, although not as realistic.

number of inputs This is a very important feature with A/V receivers because if there aren't enough inputs to handle all of your different pieces of audio equipment (cassette deck, CD player, VCR, DVD player, TV, etc.), you will have to constantly unplug and replug the patch cords.

video inputs Receivers from just 10 years ago contained no video inputs and just a few audio inputs. Now, the A/V receiver is the control box for your video system, as well as for your audio system. Likewise, with all of the different video components available, it is important not just to have two separate video inputs, but to have a few for whatever equipment might be in your future.

S-VHS video inputs/outputs If you are using S-VHS equipment, look for an A/V receiver with both regular and S-VHS video inputs and outputs.

speakers The "speaker" controls on an amplifier are typically a pair of pushbuttons or switches that allow you to use two different speakers, independently or simultaneously (Fig. 5-15). Either you can use two different speakers in tandem for each channel, use different speakers when listening to different types of music, or

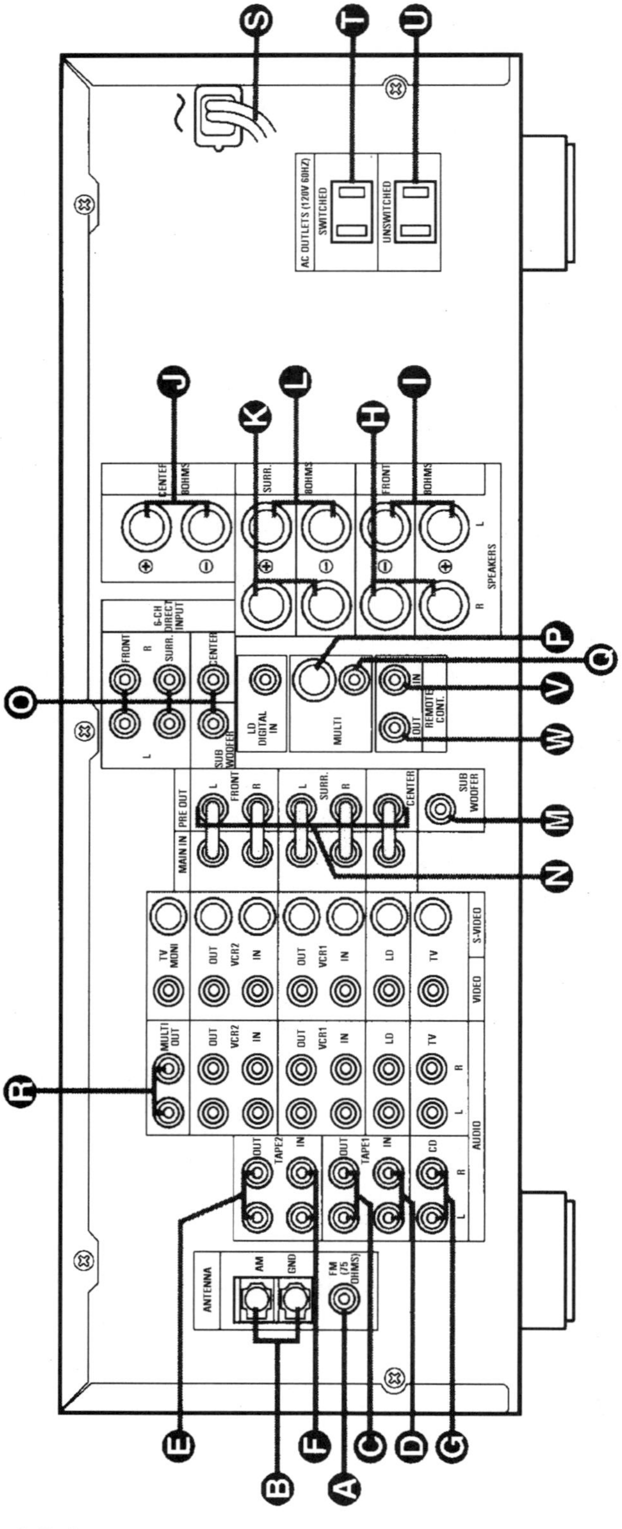

Fig. 5-14 The back of the Harman/Kardon AVR80II A/V receiver, showing the different jacks: A. VCR 1 audio in, B. VCR 1 audio out, C. VCR 1 video in, D. VCR 1 video out, E. VCR 1 S video out, F. VCR 1 S video in, G. VCR 2 audio out, H. VCR 2 audio in, I. VCR 2 audio out, J. VCR 2 video in, K. VCR 2 S video out, L. VCR 2 S video in, M. TV audio in, N. TV video in, O. TV S video in, P. LD audio in, Q. LD video in, R. LD S video in, S. LD digital in, T. TV monitor video out, U. TV monitor S video out. Harman/Kardon

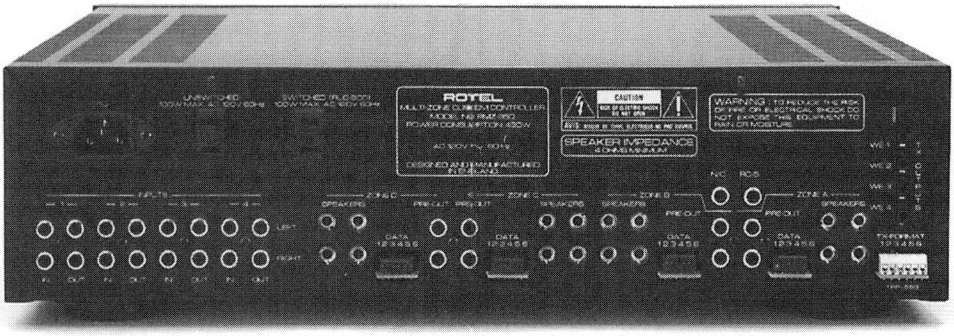

Fig. 5-15 The back of an A/V amplifier, with outputs specifically designed to power speakers in a number of different rooms throughout the house. Rotel

place a pair of speakers in front and behind you for a pseudo-surround sound.

GUI Over the past decade, computer technology has emerged in many different products. As the lines between some audio components have melted into those of the video sector, the use of computer technology in audio components has risen. The most common is *GUI (graphical user interface)*, which is also known by some other names, such as *on-screen control*, etc. With many of these systems, the user-control menus are displayed on the TV and you can input commands via the remote control. In the case of the Kenwood Stage 3 controller, the display is an LCD screen that is located in the remote control (Fig. 5-16).

RDS *RDS (Radio Display System)* is a system where radio data can be received on a standard radio. Instead of merely transmitting music and talk, some FM radio stations are beginning to broadcast data as well. Even though this signal is being transmitted along with the broadcast material, you can't hear it. Instead, this data can be received by your radio and the computer technology inside is capable of deciphering and displaying this information. None of this information is very complicated (they aren't transmitting books, line by line), but it could be extremely useful.

automatic radio muting and control circuits Some companies have added various automatic radio muting and control circuits to alter

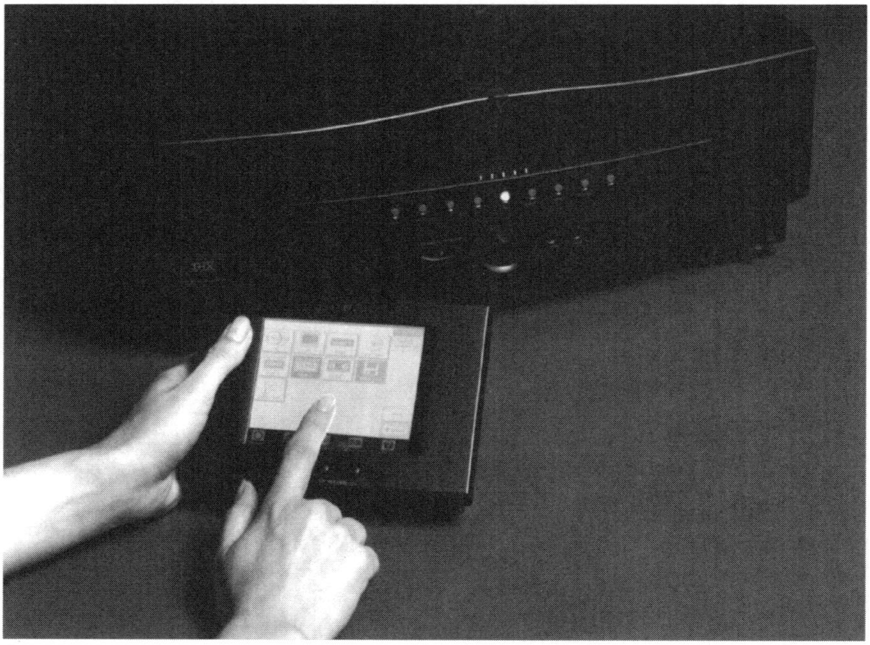

Fig. 5-16 A bit different than standard GUI, the Kenwood Stage 3 controller uses a touch screen built into the remote control, as opposed to control via an on-screen menu. Kenwood

the reception quality of broadcast stations. With some radios, if you tune off a station or if two stations are overlapping each other, the radio will play static or interference. In other cases, receivers have extra circuitry so that if a weak signal or no signal is detected, this circuitry will automatically reduce the volume of the radio so that you aren't blasted with static. A more basic feature is the automatic stereo/mono switch, which some receivers have. When a stereo signal is received that is above a certain level, it is put out in stereo; if it is below that predetermined level, it will be heard in mono so that the extra noise will not be heard. In another system, the higher frequencies are gradually rolled off as the radio signal strength decreases so that the interference will be less noticeable.

LCD display Although it isn't a function in and of itself, the LCD (liquid crystal display) is necessary for all of the feature-laden receivers and receiver/amplifiers. The LCD can be built for much more specific and complicated applications than LED (light-emitting diodes) displays can. As a result, for complicated A/V receivers, LCD displays are necessary.

preset memories One of the most convenient applications of computer technology that has been applied to radio is preset memories. In the analog era, radios had several pushbuttons. With some work, you could program each one to move the dial to a certain point whenever you pushed the button. Now, you can program a number of different buttons quickly and each will immediately put you on the exact frequency (the old mechanical pushbuttons often varied in frequency each time you pressed the button).

scanning With the scanning function, the receiver will tune through the radio band that you have selected and it will allow you to hear the different strong radio stations. Like many of the features here, the scanning function varies a bit among different manufacturers. In some cases, it finds a strong station, stops for a brief amount of time to allow you to hear what's on, and then it moves on. When it hits the end of that particular broadcast band, it stops. In other cases, the scanning function will stop every time that it finds a strong station. It will only be reactivated if you press the button again. This function is sometimes used in conjunction with a Set button.

remote control Remote control is one of the highlights of A/V receivers. With it, you can control nearly all aspects of your audio system without even having to get out of your chair. Of course, for the terminally lethargic, remote control can be more damaging than healthy.

video source-selection buttons Every A/V receiver has a number of buttons that are marked with such things as "VCR," "TV," "LD," "AUX," etc. If you press one of these buttons, the receiver will run that particular source for the video to the TV screen. So, whether you have a VCR playing, the TV on, and a laserdisc playing at the same time, if you press the VCR button on the receiver, the VCR's picture and audio will play.

Audio source-selection buttons Every A/V receiver has a number of buttons that are marked with such things as "AM/FM," "CD," "Tape," "AUX," etc. If you press one of these buttons, the receiver will run that particular source for the audio to the speakers. So, whether you have a cassette playing, the radio on, and a CD playing at the same time, if you press the CD button on the receiver, the CD's audio will play.

On-screen display In addition to the GUI programming, some A/V receivers allow you to press the On-screen display button so that a status report can be posted.

Delay On some A/V receivers, you can add a delay to the signals that travel to the rear speakers. With the delay, the room will artificially sound larger.

Mute Not a really noteworthy feature, mute allows you to turn off the sound with the push of a button.

Clear The Clear function just eliminates any information that you begin to key in.

DSP *DSP (digital signal processing)* is a process by which the signal is digitized and altered. In some other technologies, DSP is used to eliminate noise or even for artificial speech production. In A/V receivers, DSP is used primarily to manipulate the audio so that it sounds like it's being played in rooms of different sizes. Most A/V DSP functions consist of settings like "Concert hall," "Auditorium," "Blues hall," and others. The effect can be kind of nice for listening to music, but it's not as handy for the audio tracks of video: Why would you want to make some intimate dialog sound like it's in an auditorium?

Preamplifier outputs The essential function of an A/V receiver is as a control box. Because A/V receivers emphasize their functions, not their amplifying characteristics, you can imagine that some of this equipment isn't as clean-sounding as some audio-only amplifiers. Another possibility is that although the equipment sounds good, it just isn't powerful enough. If the A/V receiver has preamplifier outputs, then you can use the A/V receiver as a control box/ preamplifer and a power amplifier to boost the audio to an acceptable volume.

PUTTING TOGETHER THE AUDIO SYSTEM

Now, it's finally time to put that audio system together. You have a TV, a VCR, and maybe a laserdisc or DVD player. Perhaps you've been playing all of the audio back through the speaker(s) in your TV cabinet. What features do you want and what system do you purchase?

First, you need to look at the amount of money that you have available for an audio system. If you are on a low budget, but would still like to make the ole TV sound better, just use a stereo receiver and speakers to improve the audio. It doesn't take much to make a mono TV sound better. Even a mediocre old amplifier and speakers will significantly improve the sound. Best of all, plenty of good, well-made old audio equipment is still out there. At the last hamfest that I attended, I bought an absolutely pristine, beautiful Pioneer receiver from the 1970s and an old Panasonic cassette deck in very nice condition for $40. With a deck like that and a few speakers from an old stereo system, you could put together a good-sounding mini home theater that even looks nice for less than $100. The main thing that you would be lacking is the convenience of being able to remote-control everything.

To add more audio flexibility, you could buy a used audio mixer (such as the little units manufactured for Radio Shack or by Shure). Just be sure that the units can mix line-level audio, not just microphone audio. With a typical audio mixer, you can easily mix or fade between four different audio sources. With some searching, you could still find a nice amplifier, speakers, and mixer for less than $100. You won't be able to quickly switch between video sources, but that's a tradeoff of low cost.

To improve this basic system, you could add another set of speakers, if the amplifier has outputs for four separate speakers. You could place the second set of speakers behind you so that you have a form of surround sound. Aside from these changes, there is little else that you can do without replacing some of the system components.

The next big improvement that you could make would be to replace the old audio receiver and mixer with a good A/V receiver. If you can't quite afford a good A/V receiver, don't settle for one with a few less inputs or one that seems to be of lesser quality. Hold on until you can get a model that suits your needs. Settling for an intermediate unit could waste quite a bit of your money. As is the case when looking for any equipment, be sure that it is manufactured well. Look for good-sounding audio, S-VHS inputs, long-term warranties (anything over one year), and outputs for amplifiers and subwoofers, rather than flashy features (such as DSP) or do-nothing features (such as mute).

If you take the step of buying an A/V receiver, you can always add a power amplifier (to really boost the signal) or subwoofers (to make everything sound more life-like).

CONCLUSION

This chapter is just intended to provide a little information on the audio aspect of your home theater system. Plenty of more serious information on audio systems is available. For example, see *Home Audio* by Andrew Yoder.

6 A/V System Wiring

The most forgotten part of any audio system is what actually transmits the power, video, and audio signals from one part of your system to another. Wire and cabling are often forgotten because they are about as boring as a Donny Osmond record. Why would you waste your time and money working with rolls of wire or cabling when you could purchase and play with some really cool equipment? I have no idea. I would much rather play with all of the buttons and displays!

When you get right down to it, the subject of wire and cabling is very important. With a bad choice in wire, you could decrease the quality of the sound or pickup interference. Although it is unlikely that a poor choice in wire will cause a house fire or an electrical hazard, the difference is in efficiency and overall picture, sound, and recording quality. This chapter covers topics that are wire and transmission related, such as the wire itself (including wireless systems) and installing ground systems.

Wire

Other than one peculiar exception, wire or cable (often coaxial cable) is always used to transfer audio and electricity between the various components. Few people think about the different qualities of wire and the efficiency as a signal travels from one component to another. However, wire is made from several different materials (of many different qualities) using different drawing methods. Also, wire can be either made with one single wire, with many strands, or with bundles of strands; it can be shielded to prevent other signals from causing interference; and multiple wires can be encased within either a round or a flat cable (Fig. 6-1).

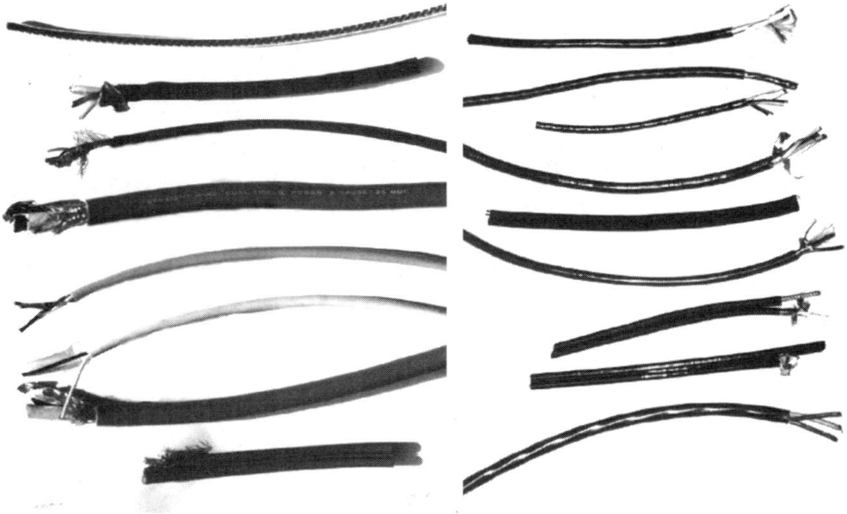

Fig. 6-1 A number of different types of speaker, interconnect, and power cables.

Most audio, video, and power cables are made from copper. But if you use wire that was intended for some other electrical application, be aware that it might be made from another element or elemental combination, such as aluminum or copper-coated steel. Generally, copper is the only element to use to transmit audio and power signals through your house. Aluminum is fine for use in electric fences or in long radio antennas, where low cost is more important than high efficiency.

As stated in the previous paragraph, copper is an excellent conductor of electricity. Several other elements, such as silver and gold, conduct better. However, their costs are obviously prohibitive for most electrical applications. Silver is occasionally used, though, for very high-end wiring applications; gold is commonly plated over wiring connectors (such as the inputs and outputs for VCRs, DVDs, amplifiers, speakers, etc.).

WIRE SHAPES AND STYLES

A number of problems can occur when different components are connected together via wire. Two major problems are resistance and skin effect. Skin effect is the scientific property by which electricity travels on the outside (the "skin") of a conductor (Fig. 6-2). It has nothing to do with rashes, blushing, acne, leprosy, or a host of other dermatological problems. Skin effect causes different problems in different applications. For example, when transmitting at radio frequencies, if a station is outputting too much power into too small of a conductor (one that has too little surface area per watt of power), the resulting heat could cause the radio antenna to burn up. However, at audio frequencies, skin effect is a completely different monster. Instead of the "bigger is better" maxim that rules for radio conductors, larger-diameter wires can cause other problems that result in high-frequency distortion. This distortion problem occurs in both solid and multistrand conductors.

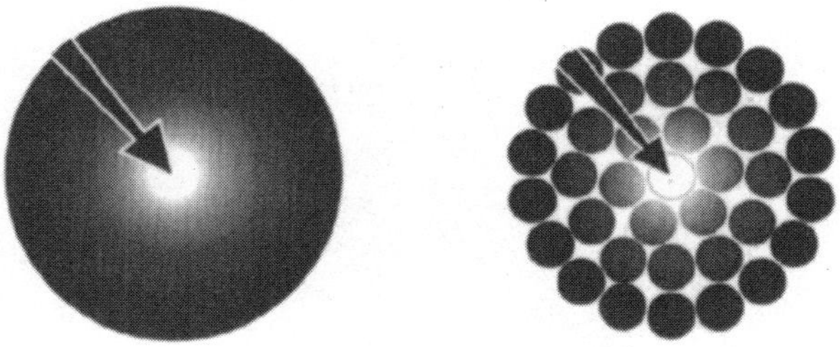

Fig. 6-2 A visual representation of skin effect. © AudioQuest

Resistance is another problem that plagues electrical circuits. Resistance is the opposition to an electrical signal flow. Although the resistance through a wire will cause a slight decrease in the signal strength, it will also cause some distortion. So, high-resistance wire might be great to make the coils in toasters or space heaters, you want you find very low resistance wire for the speakers and interconnections in your home theater system.

Solid (single-strand) wire has a much higher resistance than stranded wire. You can see that problems occur with this type. If the diameter of solid wire is too large, the signals will distort as a result of skin effect. Yet, if the diameter is too small, the resistance of the wire will cause distortion. Worse yet, the two boundaries overlap and you can be stuck with some sort of distortion, no matter what the wire gauge is.

Now that your dreams of using solid-conductor house wiring for your audio system have been dashed, it's time to move on to the different types of stranded wires. The typical speaker wire is usually something along the line of "zipcord," which consists of two insulated parallel conductors of stranded wire. Unlike parallel twisted hookup wire, the two independent zipcord wires are connected along a seam that runs the length of the wire.

Returning to the problem of skin effect in speaker cables, notice that some companies try to reduce this dilemma by gradually turning the wires throughout the length of the cable (Fig. 6-3). In

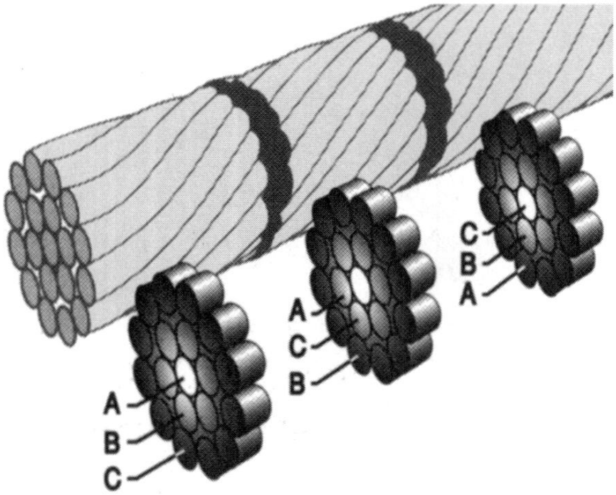

Fig. 6-3 Separate wire strands running throughout a cable length. Notice how the strands shift position in the cable to reduce skin effect. © AudioQuest

the process, wires that are in the center at one point are later shifted to the outside, where most of the current is being transmitted. This technique greatly reduces the negative results of skin effect. To further eliminate the negative effects, sometimes either stranded bundles or solid wires are gradually turned around an insulating core. As a result, each of the wire strands within the bundles or surfaces of the single-strand wires is being slowly turned, so the anti-skin effect is doubled. Sometimes, these solid wires or bundles are insulated and kept a certain distance apart to prevent them from interacting (Fig. 6-4).

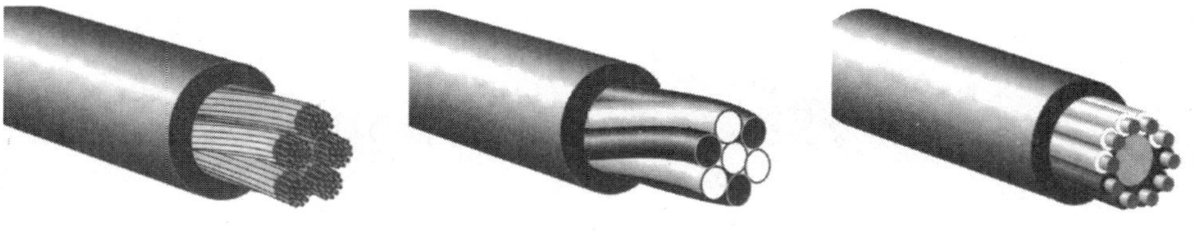

Fig. 6-4 Insulated wire bundles. © AudioQuest

Another type of audio cable uses several parallel solid conductors or stranded wire bundles (Fig. 6-5). The purpose is much the same as the theory behind the bundled wires from the last paragraph. Like the other types of wire, the parallel multiconductor ribbons are made with a variety of wire types, spacing distances, and jacket and insulating materials.

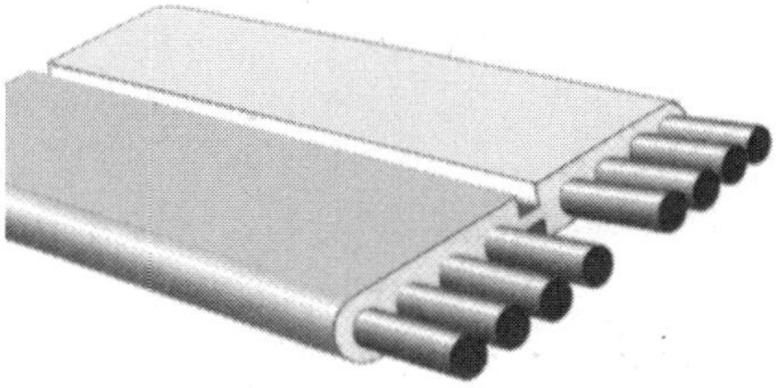

Fig. 6-5 Parallel flat cable.
© AudioQuest

The last general category of A/V wire is that of shielded cable. Shielded cable is any cable that shields the wires inside and prevents various types of electrical interference from entering the lines. Shielded cable is especially important if your electrical system or appliances are noisy or if the speaker cables will be running parallel to power cables. Likewise, some power cabling is also shielded to prevent electrical impulses from emanating out from the cables.

One of the most common forms of shielded cable is coaxial cable, which consists of a center conductor that is surrounded by an insulating dielectric material. A layer of braided copper wire shielding wraps around the dielectric (it makes a sort of cylindrical copper plaid pattern), and it is covered with an outer insulating jacket, which consists of a hard, rubber-type material (Fig. 6-6).

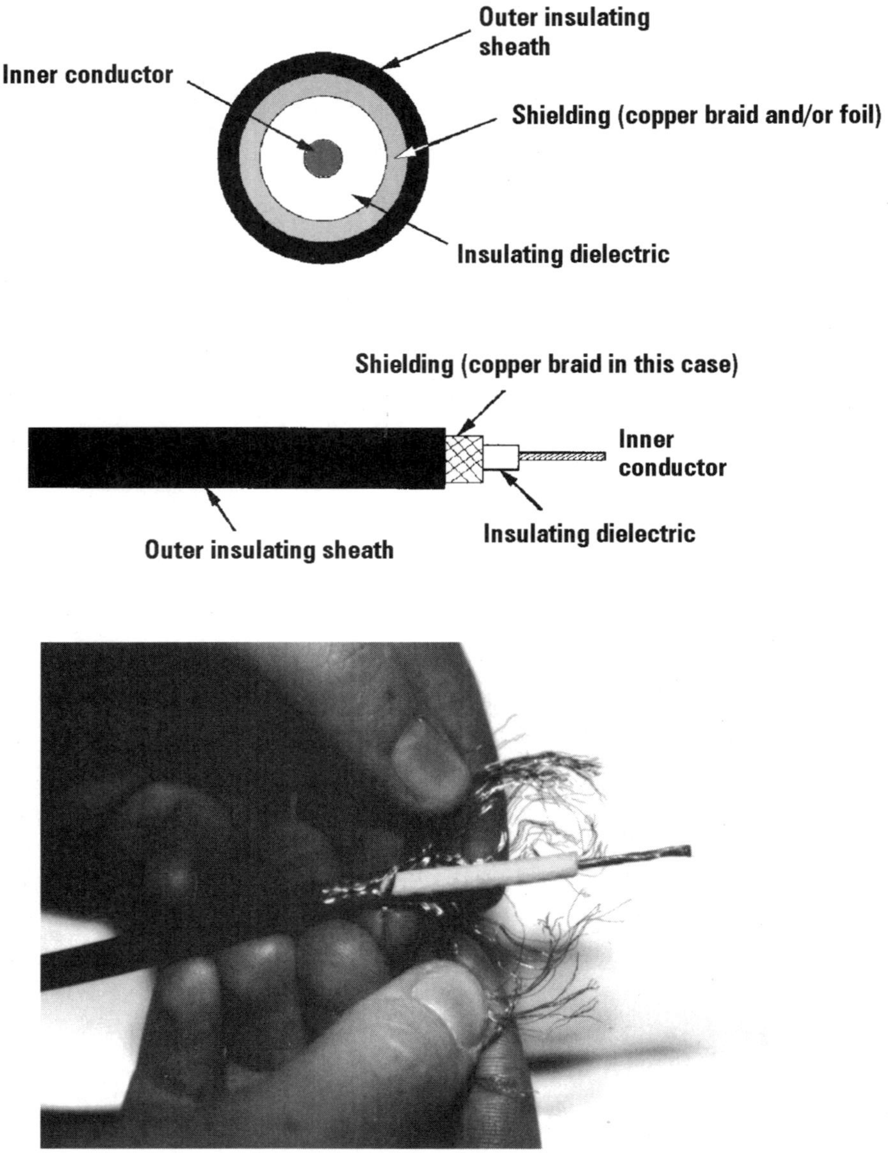

Fig. 6-6 A breakdown of coaxial cable, showing the separate parts: End view (top), side view (middle), and with the copper braid pulled apart (bottom).

One of the most visible applications of standard coaxial cable is for cable television. Coaxial cable is always used to route the television signals from the source to the home (at least in all of the television cable runs that I've seen). A number of different types of coaxial cable have also been designed for audio-specific purposes (Fig. 6-7).

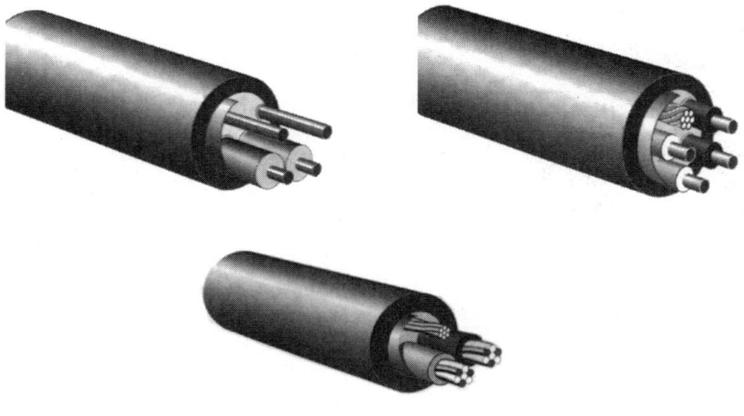

Fig. 6-7 Several different types of coaxial cable that are intended for audio purposes. © AudioQuest

If you are installing long cable runs throughout your house, you should consider shielded cable. Long cable runs are susceptible to picking up noise from parallel runs of electrical wires (known as *crosstalk)* or from other devices. Crosstalk was discovered in the 19th century, during the early days of the telephone. Long parallel lengths of telephone wire would sometimes receive audio from the opposite wire. Shielded cable will typically eliminate this type of interference.

So, which cable is best for your home theater system? In general, the aforementioned cables are all of much better quality than those that you would typically find packed along with a box of speakers or that you could purchase from your local electronics store. Most of the specialty cable companies specifically produce their cabling for audiophiles and video fanatics (such as AudioQuest, Straight Wire, Kimber, Tara Labs, etc.), so their products won't be low grade. The differences between the specialty cables probably won't be important to anyone except audiophiles.

SHOULD I PURCHASE HIGH-END CABLES?

The science of cabling for home theater is quite esoteric and it is one of the most argued-over topics in the controversial A/V world. Claims about the cabling vary greatly, from being an audio cure-all to being an absurd case of snake oil advertising. I won't tell you to spend a large sum of money on high-end cables, but I won't say that these cables won't help in some specific applications.

The problem is that when the cables use precious metals, for example, they are bound to be hugely expensive. And if you can't afford a few thousand dollars for your stereo system, why would you blow several thousand on the audio, video, and power cables? Many of the high-end wire companies rely on two factors (proverbs?) to sell their products:

1. More expensive is better
2. A fool and his money are soon parted.

So, is it foolish to spend your money on expensive cables? Not necessarily, but that depends largely on how much money you have and are willing to spend on your home theater system. And how much is expensive? Well, Transparent Audio sells their MusicWave Reference cable for a list price of $4100 per 8-foot pair! It would be foolish for me to outfit a system with this cable because it would quickly be repossessed, along with my house and cars! It wouldn't be foolish for someone who could easily afford a $200,000 audio system. I'd say that those people could find a more-effective use for their money, but that's just my personal view.

Like everything with high-end electronics, if it is important for you to have the best and you have the money, then buy it. Otherwise, if the different grades of cables are a concern to you, then do some comparison shopping. Check the cables in different systems if you have a chance. Or, talk to some salespeople at different local stereo installation shops and see if you can do some comparative listening and viewing. If you are lucky, you might be able to test the same system using different cables. Of course, beware of some of the sales tricks, such as turning the volume up higher on the one amplifier so that the system will "sound better." If you have some friends that have trustworthy opinions on home theater, then you should talk with them if you can't find any cabling to use as a reference.

After these tests, choose the cabling that sounds or looks the best for the lowest price. Even if you can notice a difference, the better cabling might not be worth extra cost.

Also, consider the equipment that you have and that which you hear the cable being auditioned with. What grades of components are used in your current equipment? Unless silver and gold are used throughout your amplifier (i.e., if it originally cost $3000, there's no reason to waste your money on silver wire). For that matter, Radio Shack's cheapo ($3.99 for 60 feet) was rated highly in the latest "Recommended Components" edition of *Stereophile*. If the differences are negligible, I wouldn't even pay $10 per foot (relatively inexpensive in the audiophile market) for better-grade wire.

Finally, if you are desperate for something decent at a good price, go to a few local hamfests (as mentioned in further detail in Chapter 1). I've found some really great deals on partially used wire spools. Two years ago, I bought a new 500-foot spool of #14 stranded copper wire with a plastic-covered rubber coating for $12. A few months earlier, I found a mostly used spool of silver coil wire. It is finely stranded #18 silver (or silver alloy) wire with a Teflon coating that was probably used for coils in high-voltage applications, where other insulations would have burned off. My spool, with approximately 200 feet of wire, cost $4! This stuff wasn't originally sold for $100 per foot new, but it certainly wasn't cheap! This Teflon/silver wire isn't intended for home theater applications, but I have also seen plenty of high-grade, inexpensive patch cables and cable harnesses available at hamfests.

FM AND TV TRANSMITTERS

The one peculiar exception to using wire for transporting signals from one location to another is practical for large houses or for just the convenience of transporting a signal from one room to another. Or you might want to transport the signals from your VCR to TVs throughout your house or office building.

A number of problems plague long cable runs. The potential for interference grows significantly as the wire length increases, the signals passing through the wire can weaken (important for line-level audio), and the costs can be prohibitive. If you can avoid using wire, why not?

The easiest way to link audio from one source to another is via a radio link. You get a transmitter and transmit the audio to a receiver in a different location. Easy! This concept has been used for years for rock bands, who were tired of tripping over everyone's audio cables. They just went "wireless."

Guitar transmitters operate on special radio frequencies, outside of the broadcast bands, to minimize interference. Because they were designed for the professional application (playing electric guitar or bass in a band), the price of this equipment is high. For that matter, the transmitting frequency isn't really appropriate—it would be much easier to use a transmitter that operates on a standard broadcast frequency so that you could use your stereo system or a portable system to receive the signal.

Back in the 1960s, a few companies realized the usefulness of very low-power transmitters for the broadcast bands. A few companies, such as Knight, built little "phono oscillators"—tiny transmitters that were intended to carry audio from your turntable to any broadcast radio in your house. These little transmitters were originally sold as kits, with one or two tubes, for operation in the AM broadcast band. These units evolved over the years, switching over to FM and transistors, and the applications changed, as well.

Today's versions of the old phono oscillator are tiny little FM transmitters that are typically intended as "home broadcasters" or bugging devices. Some people even broadcast to their neighborhoods with them, but those applications, along with bugs, are well beyond the scope of this book. Aside from the typical kits that are available for these applications, a number of different manufacturers are now offering small FM transmitters (called *FM modulators*) along with their in-car compact disc changers. In general, these units are only found in the higher-priced changers. However, Sound Barrier does offer the CD-10FMC 10-disc changer for a list price of $399, which is considerably less than the other models with FM transmitters.

If you are interested in purchasing one or more little FM transmitters for use around the house, be sure to ask for recommendations. Some of the transmitter kits are very unstable (the frequency drifts), so you will have to constantly retune your receiver. Others have limited frequency response, so the highs and lows might be lacking. Still others have poor stereo separation. Your best bet is finding a decent "FM modulator" from a car stereo dealer or mail-order store. These units are typically made specifically for good audio and frequency stability.

I doubt that anyone will make this mistake, but be sure to use an FM-band mini transmitter, not one constructed for some other frequency range. Not only will the audio likely be poorer, but the results are less predictable. For example, a few years ago, an amateur radio operator in North Carolina decided to build a mini shortwave transmitter to transmit the local FM rock station from his house to his workshop. His transmitter was a tiny 5 watts, but it was powerful enough to be illegal to use. Before long, agents from the Federal Communications Commission arrived at this man's house to close the station—it had been heard as far north as Boston! And these distances weren't a fluke; I heard the station for hours at a time while listening in Pennsylvania.

Most of this section has covered little FM radio transmitters. Although little TV transmitters are less common, they are available (Fig. 6-8). Most of these are for amateur TV (ATV) applications (which use different frequencies than standard TV broadcasting, so beware!) or for local hobby broadcasting. If you don't plan to broadcast your interpretive dance version of the local news to your local community, then you might want to try a different trick. Buy a wideband amplifier for receiving purposes from an electronics store, such as Radio Shack. Then, connect the output signal from the VCR to the input of the amplifier. Connect the amplifier to a small antenna (perhaps rabbit ears) and away you go! It won't output much power, but it should provide good reception throughout a few rooms of your house.

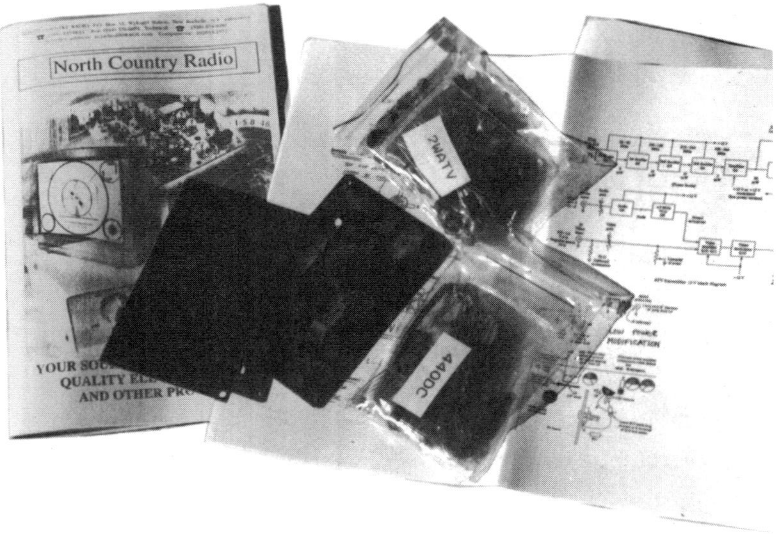

Fig. 6-8 Why keep a VCR, laserdisc, or DVD player in every room, when you can just broadcast the signal from a TV transmitter? Shown is an ATV transmitter and receiver kit from North Country Radio.

Another possibility is the cable TV transmitter. A friend of mine picked one up for a mere $10 at a recent hamfest. These units are top-grade, but it's rare to find one that is so inexpensive.

Now, a warning. Be sure that the signal isn't traveling outside of your property. If it does, particularly if you connect this system to an outdoor antenna, you could be emitting too powerful of a signal and you could be in violation of FCC (in the United States) or DOC (in Canada) rules.

MAKING CONNECTIONS

Making good connections is quite important if you want your system to look and sound as clean as possible. High-end audio connections are made in a of a bit different manner than connections for nearly any other electrical application. Normally, either the wires are "tied off" together or they are soldered and possibly weatherproofed. Some solder and a few wire nuts would be all that you would need to install the wiring on most electrical systems.

However, this is a far cry from connections in the A/V world. If you want to see some looks of terror and/or absolute disgust, talk to a professional installer about how you connected your system of equipment from Audio Research, Martin-Logan, and Parasound using only 18 wire nuts! That's worse than ordering caviar from an exclusive French restaurant and dousing it with cheap ketchup.

Fortunately, you shouldn't have to make any splices in your A/V lines. You should be able to run a wire straight from one audio or video component to another (Fig. 6-9). The only extra work that

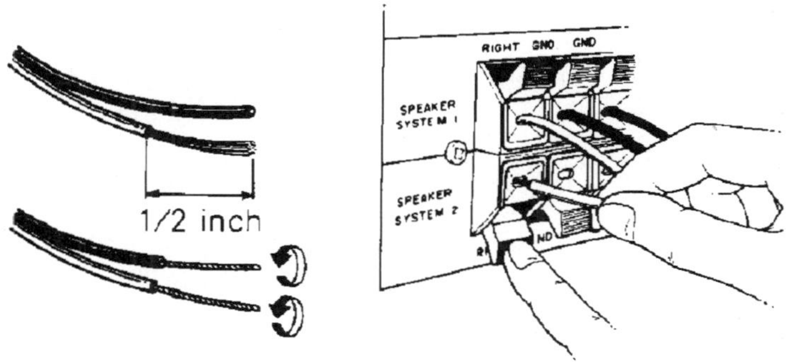

Fig. 6-9 Preparing speaker wires for connections in push-block terminals.
Harman/Kardon

you might have is with soldering connectors onto the ends. Although many of the connectors are so-called "crimp-on, no-solder connectors," most require solder or the joints will become electrically intermittent.

Remember that solder should only be intended as an electrical connection, not a mechanical one. Although solder will strengthen any joint, it does not handle physical stress very well. Some audiophiles will say that solder is not much of an electrical connection either. The antisolder folks claim that because solder is a mixture of lead and tin (neither of which conduct electricity nearly as well as copper), the signal loss and distortion will be severe at these joints.This is a valid point when putting together a high-end A/V system. However, an audiophile would only use single runs of cable from one component to another. If you are even considering using spliced wire, you probably will not be bothered by a slight audio or video degradation, resulting from using a soldered joint. If this is the route that you want to take, take your time and solidly crimp the wires together before soldering so that the wires still make very close contact with each other. The solder would then serve to make the joint a bit stronger (both electrically and physically), and it would also help to waterproof the joint. See Fig. 6-10 for a look at the soldering technique.

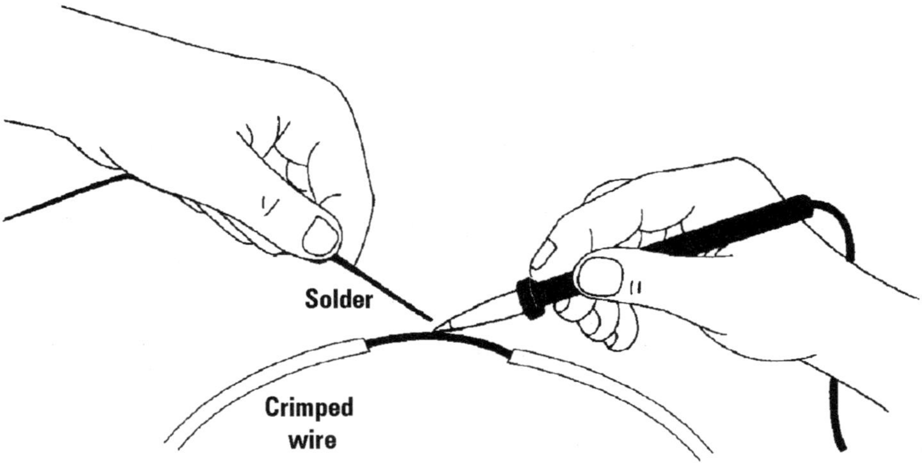

Fig. 6-10 Making a solder joint.

Wire nuts are a poor method to connect the wires together because the joints are not especially reliable. A few years of vibrations, being moved to accommodate the vacuum cleaner, etc., will cause these connections to become even less reliable. Most people who connect wires in home theater systems use these splices because their wires are too short to reach from the speakers to the amplifier. If this is the case, just get the correct length of wire so that you can avoid future problems with your system.

CONNECTORS

Choosing the proper connectors for your system isn't a big deal, but the subject does bother a few people who are trying to eke out the most from their stereo. The concern about the connectors is that the metal might be of poor quality. Better grades of metal have a higher conductivity, and the better the conductivity, the higher the percentage of signal that will be passed from point A to point B. As a result, most audiophiles only use connectors that have been plated with precious metals, typically gold, but occasionally silver. Gold is most often used for A/V applications, and silver is typically used for radio (antenna connectors, etc.). Choosing connector metals is like choosing cables, if your equipment is really high end and uses silver throughout, go with gold connectors. If not, skip the unnecessary extra cost.

The most important concern for audio enthusiasts, with respect to connectors, is that they are built tough. The really cheap RCA plugs use a thin metal (that I assume has inferior conductivity) with a small clip-on plastic protective cover. These plugs can easily be ripped right from the wire or become intermittent. The next-better grade of plugs is solid enough to survive years of use in a system. These plugs use stronger metal parts and the plastic or bakelite-type protective cover screws onto the plug. This type of plug is also available with gold plating on the conductor. Don't worry about purchasing the real high-end-style of plugs. These typically use a metal screw-on cover (this much isn't overkill) and a metal spring to wrap around and protect the wire from breaking inside the cord. These plugs are intended for musicians or recording studios, where quality metals are necessary in an interconnect that is constantly plugged in and pulled back out. You won't be pulling the plug in and out nearly this much, so save a few bucks and opt for something a bit less tough.

PLANNING AND ASSEMBLING THE SYSTEM

As stated earlier in this chapter, the wiring in a home theater system is one of the most forgotten aspects of an installation. In terms of being overlooked, wire rates with the spider in a speaker.

When you are first assembling your A/V system, the wiring goes unnoticed. But when the need to rewire occurs, then it's time to scratch your head and wonder why you didn't think about this earlier. Depending on when you are running the wiring, it can be anything from very easy to extremely difficult to install or replace.

Installing or replacing the speaker wiring can be a major task. Before you start to install the system, you should determine what type of cable you will be using and how much you should purchase. The system planning turns into a real mess if you have a TV/video system, a separate hi-fi system, and a multimedia computer system somewhere else and you decide to interconnect them all. For convenience sake, you should try to set them all up as close as possible. For example, if you plan to try your hand at computerized digital video editing, it is not convenient to record and play back between your computer and VCR if both are contained in separate rooms. If possible, it might be best to either work from a laptop computer by your integrated home theater system or you might rather have a little computer suite nearby, for example.

To find the lengths, determine where the speakers and the stereo system will be positioned. Then, use lengths of string or wire to find the exact lengths. Be sure to account for any extra bends or corners that the wire will be making. Allow an extra 6" to 1' (15.24 cm to 30.48 cm) for every long cable run and about an extra 3" to 6" (7.62 cm to 15.24 cm) to be safe. It might seem like a lot of extra work just to measure the wires, but when you deal with custom wiring, that extra foot or two here and there could add up to quite a bit of money. Then, make a note of any specific connectors that you might need, such as RCA plugs (Fig. 6-11), slide-on speaker clips, etc.

Fig. 6-11 StreetWires cables with specialty RCA plugs at the ends. StreetWires by Esoteric Audio USA

When you plan out the wiring for your system, try to run the audio and video cables away from the power wiring in your house, or at least use shielded cables to prevent noise from entering the audio cables. Even if you can't place the power and the A/V cables on opposite ends of the walls (top and bottom, or on either side), at least try to space them a few inches apart. This distance alone will decrease the noise pickup. Before you start running the wires, be sure that you have made any solder joints that are necessary.

The easiest situation would be in a new house that is in the early to mid-stages of construction. Here, the studded walls would be up and the roof would be finished, but the drywall would not be installed yet. So long as you have accurately determined where you are going to place the speakers and the audio equipment, the rest is easy. First, you would mark the places on the floor where you expect the equipment to be placed. Then, drill holes through the 2-X-4 studs for the wires to pass through. If your listening room is on the top story of the house, it's best to place all of the horizontal wiring runs above the wall so that you can easily access it through the attic or crawl space.

Some people take pictures or videos of the house while the beams and studs are all within view. This way, they can get back into the walls without cutting into electrical wires or nails. A really great method of viewing the house is to take slides of the walls while the house is being constructed. If you need to get back into the walls, you can project the "skeleton" of the room up on the walls. Neat!

The photo system of wire tracing can be very helpful, but it's not the easiest method. But merely discovering where the wires are placed in the walls doesn't make pulling the wire out any easier. Heavy gauges of wire will often catch on the edges of holes and pull fast, like a knot. To prevent your cable from knotting up in the walls, you might consider drilling holes in the studs, then passing small-diameter PVC pipe through the holes (Fig. 6-12). Rather

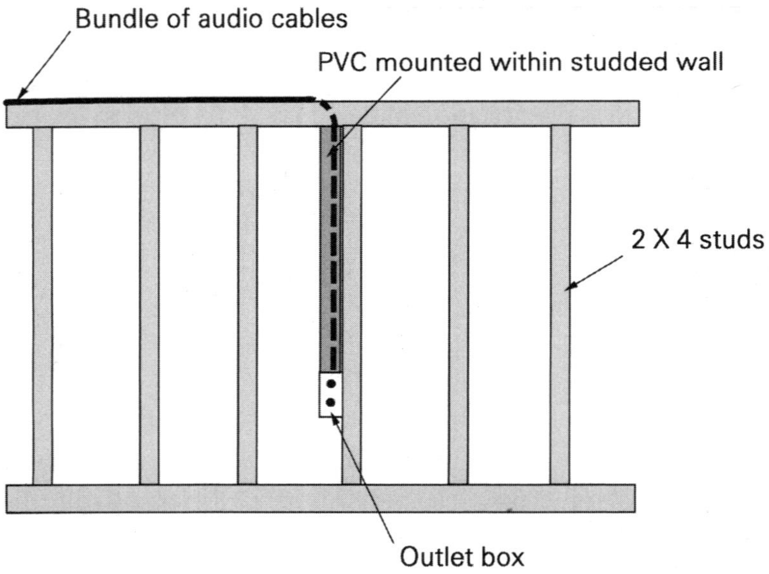

Fig. 6-12 Running audio cable bundles through studded walls.

than just run the wires out through a hole in the floor or walls, use a plastic electrical junction box. Cut a hole in the junction box so that an elbow joint for the PVC pipe can fit into the junction box. Mount the junction box on a stud, put an elbow joint on the pipe, then fit the other end of the joint into the junction box. Now you can easily feed the wire through the pipe, from one junction box to the other. You can either mount jacks onto the junction box plate cover or you can run the wires directly through the cover to the speaker or equipment (whichever is the case).

The real problems occur when you start trying to install system wiring in an already-constructed house. You wind up trying to drill holes in walls, ceilings, and floors, and getting wire stuck or lost. If you have a typical system, you will probably either want to run the audio cables along the baseboard or under the carpet at the edge of the wall. If you choose the former method, you will probably want to staple or tack the cord along the baseboard. Just be sure that the metal doesn't cut or squash the wires. If you go for the under-the-carpet method, you'll have to pull up the quarterround around the baseboard and slowly work up the carpet around the edges. Then, run the wire along the baseboard, put the carpet back in place, and nail the quarterround back. Be very careful when nailing so that you don't hit the wiring.

It takes a lot of work to try to run wiring through an old house, without running it along the baseboard or under the carpet. Part of the problem is that you can never be sure what lies behind that drywall. I have friends who live in an old two-story brick house that was built in the country somewhere around 1900. All of the walls that directly face the outside are brick, covered with plaster. Aside from studding up the inside of the walls, there is no way that you can run wires through the walls.

Even if you are lucky enough to know that the walls are studded and clear of major obstacles, you will encounter other difficulties. For example, you can't easily mount a junction box in the wall if the drywall is already in place; there is simply no space to nail it in. If you want to put one in, you will have to rip out some drywall, a piece from stud to stud. Then, you'll have to nail in the junction box, run the wires through the box, cut a new piece of drywall and nail it in, then go through the whole process of taping, applying wall compound, sanding, and painting. That's an awful lot of work just to make the stereo wiring less visible!

If you want to see some of the problems and questions faced by an audio enthusiast wiring his very large new home, visit:

http://www.mcdata.com/~meh0045/homewire/wire-guide.html

If you are building a new house and working out the details for the system, this site is a must to see.

Grounds

The ground system in a home theater system is often forgotten because not having one won't prevent you from playing back music and hearing it through your speakers. Also, although most audio equipment contains ground connections, most video equipment simply relies on the third prong of the power connector for the ground. Because you have to bury metal in the ground to create a good ground, the subject has attained an almost New Age kind of aura. "You bury stuff in your backyard to improve your stereo performance, eh? (aside) Call the cops, quick!"

Part of the reason that you won't hear much about the ground system for electronic equipment is because good grounds are relatively inexpensive to make—they depend more on getting some metal into the dirt, and less on buying a piece of expensive equipment. So, with little money involved, neither the equipment manufacturers nor the magazines will commercially benefit from writing about hi-fi ground systems.

Grounding equipment is important because cables and connectors receive electrical noise (both natural and man-made) and radio/TV signals through the air. Unless these signals are disposed of, the audio will be corrupted. The only way to dispose of these electrical signals is to dissipate them into the ground or some other large conductive object (Fig. 6-13). The best ground is the earth—especially that which consists of conductive materials (primarily water and salt). Now you need something to connect your equipment to that excellent electrical ground that you think of as a backyard.

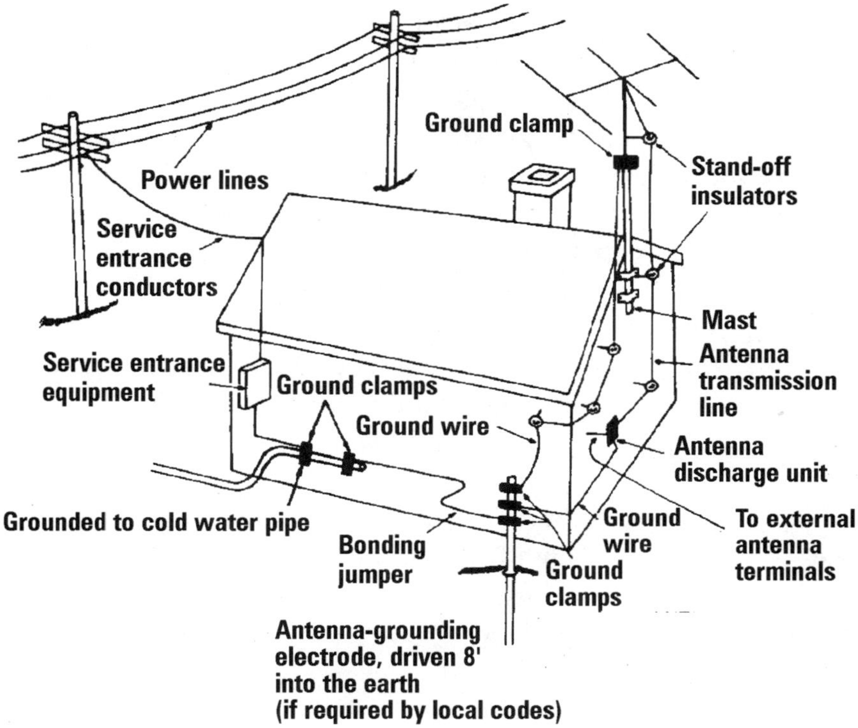

Fig. 6-13 An efficient ground system is important for good audio and video quality and electrical safety.

The best method is to connect the grounds of all of your audio equipment together with grounding wire (typically, about #12 stranded copper wire or grounding strap). A ground wire is then run from a central piece of equipment (usually, the amplifier) to the ground assembly outside. The actual ground system generally consists of very long copper or copper-coated steel ground stakes driven into the earth, a pattern of wire radials buried several inches down, or a combination of the two.

The ground stakes are available from some electrical supply stores or hardware stores. Although some of the grounds that people install for their house's electrical system are only about 1 or 2 feet long, a good grounding stake should be at least 4 feet long, and it should be pounded into the ground so that only a few inches stick out above the grass surface. I just hope that you live in an area with nice, smooth dirt. I've spent most of my life living in mountainous areas, and sledgehammering stakes through rocks and clay is a pain, to say the least. Be really careful when pounding these into the ground—they're metal and you're using a sledgehammer! You could easily slice yourself with the stake or mash something (your foot, perhaps?) with the sledgehammer. If you have any doubts about your physical capabilities or if you are accident prone, consider having a professional install the stake or try the next grounding option. Whether you install it yourself or have someone else do it for you, be sure to mark the end of the ground stake with a little flag or cover it with something so that people won't trip and fall on it.

Another factor to remember when planning your ground system is to be careful *where* you pound that stake. Remember all of those "Call before you dig" signs! You are better off calling the gas or electric company so that you can be sure that you don't drive it through one of their lines.

Rather than drive a huge metal stake in the ground, you can run a system of ground radials under the yard. The system merely consists of a number of copper wires, all of the same length, that radiate from a single point, where they are all soldered together. Eight or more radials should be fine for your ground system. Although the length of the wires is specific for radio applications (such as the ground radials for an AM broadcast transmitter), don't worry about cutting them to a specific length—anything greater than 10 feet long should be fine. The wires should all be cut and the insulation should be bared for about 4 inches at one end of each wire. Dig out paths in your yard for the radials, string the wires in place, and twist the bared ends together. Solder it and dip it into liquid rubber (used for insulating the ends of metal tools) to protect the joint from weathering and corrosion. Then, be sure that all of the wires are well covered by dirt; if they aren't, you could hit one with a lawn mower and cause some serious damage!

Conclusion

Wire, cabling, connectors, and transmission systems are not the most interesting aspects of home theater systems, but they certainly are important. They require some extra work to estimate lengths, plan, and install, so it is important to take your time and think about the arrangement.

7 Living space

Any time that you flip through a technical hobbyist magazine, whether it covers video, audio, radio, computers, or something else, the focus is the equipment. However, numerous external factors also affect how well you can either work or enjoy something. Your home theater system is no different. If it is in the wrong place and is poorly organized, you might not use it often or you might not enjoy it nearly as much.

It's kind of lame to plan out everything that you do all of the time. It's like those stereotypes of the accountants, who are so uptight about schedules and organization that they never enjoy anything. I'm not advocating any kind of devotion to the cause of organization, but if you spend some time thinking and planning about how your home theater system is organized, it might help significantly.

A good example of this is my workspace. When we moved to the house that we presently live in, I set up an office in my basement. It was perfect. I was in the basement, away from everything and everyone else. I had thousands of reference books and magazines all around me, and I had a few desks loaded with electronics and computer equipment. Although it seemed like the perfect location, I was so far removed from everyone that I wound up never working down there. Most times, I would have rather been working near my family, and anytime that I tried to work down there, I got called upon for minor emergencies anyway. With a little more planning, I came up with an audio/video/computer-integrated work area upstairs that has been fantastic. I get loads of work done and I even enjoy hanging out there. The only difference was my planning and organization.

ORGANIZATION

Chances are that your home theater system will go in the same place where your TV currently is—in the living room or den. You would probably plan to just replace the equipment where it is in your room. After you replace things for a while, you might notice that the system isn't either entirely functional or efficient. So for the best results, if you are reconfiguring your home theater system, you should rethink the entire layout of the equipment and the room in general.

If you start from the beginning, you might conceive some really great idea that would make everything go smoother. For example, you might be able to build in a false wall, sink all of the equipment into the wall and run all of the wires in the space behind it. Or, you might be able to organize everything so that you can integrate your stereo, home theater, and computer areas.

Another key organizational aspect concerns the viewing material that you have laying around. How and where do you store all of those videotapes, videodiscs, cassettes, etc.? Are they scattered all through your house in various "no-man's lands?" If so, how can you straighten up this mess and stop wasting so much time trying to find the movie that you want to watch?

The present situation at our house isn't quite a worst-case scenario, but it approaches that magnitude. First, a little background information. We have a house that's not quite a fixer-upper, but it has required complete remodeling, plus additional work here and there. It's not easy to remodel a house while working full time. So, the work has gone very slowly. At the bottom of our list of remodeling-to-do is the TV room/entertainment center area. We're not even sure where the entertainment center will go—in the living room or in an as-yet-unfinished basement room. For that matter, will we outgrow this house in a few years and decide to move? These are all very important factors to consider before spending thousands of dollars on an entertainment room.

As a result, our entertainment center plans are in limbo. We can't enact any plans to improve our current organizational disaster, yet life must go on. Our much-too-small TV cabinet contains about 50 to 70 of our most-watched videotapes. One of the favorite pastimes of our lovely 19-month-old daughter is pulling all of the videotapes onto the floor and yanking off the cardboard sleeves. This absolutely wrecks the main portion of our video collection. Most of the rest of our video collection has been sitting in a cardboard file box in the attic for a few years. With a "system" like this, we can rarely find anything quickly and many videos are very difficult to locate. Sometime, it would be nice to have the better part of one full wall loaded with video-storage cabinets. Inside could be either the slide-out or swing-out shelves.

With these videos then categorized and alphabetized, they would be very easy to locate. And better yet, we could even place child-proof latches on the cabinet doors!

Of all of the work involved with installing cabinets, remodeling a room, and picking the perfect storage device, I think that the least-likely task for me to complete is that of cataloging the videos. I just have a knack for recording cassettes and videotapes, then not properly marking them or the cases. I guess that when I'm in a hurry, I just figure that the important work (recording) is being performed and I can write down the other information later. More often than not, I don't get around to doing it.

One of my friends had a desperate case of not-write-it-downs, with literally about 100 home-recorded videotapes and no labels. The summer after he was laid off from his job, he searched through everything that was in his collection and wrote the titles on his tapes. To make things worse, he had recorded everything in the EP setting, so he often had either three movies per tape or a mixture of movies and TV shows. In addition to marking all of the titles on the tapes, he numbered each tape and bought a six-foot video rack to put them into sequential order.

One of the best ways to organize your videotapes is to always have a number of blank tapes on hand. Then you can record similar programming onto the same tape. For example, things got really messy when my friend mixed movies, Star Trek, and The Simpsons all on the same tape. He would have been better off keeping each category on separate tapes and, better yet, further breaking the movies down into categories, too.

SOFTWARE

The last step of my friend's video organization process was to catalog them all in a computer program. A handful of video-cataloging programs are currently on the market with different features. One popular shareware program is Movie Center. With it, you can catalog your videotapes and organize them according to title and category. You can also print out labels. You can download Movie Center from:

http://www.winsite.com/info/pc/win3/misc/moviectr.zip

If you want to catalog all of your audio and video software, try Media Library, available from:

ftp://ftp.cdrom.com/pub/simtelnet/win95/database/ml142sw1.zip
ftp://ftp.cdrom.com/pub/simtelnet/win95/database/ml142sw2.zip
ftp://ftp.cdrom.com/pub/simtelnet/win95/database/ml142sw3.zip

Another option is Manage Your Videos, available from:

ftp://ftp.cdrom.com/pub/simtelnet/win95/homeent/vlm32101.zip

VIDEO CABINETS

Another aspect of the organizational process that was touched upon earlier was the ability to pack everything into a place. Early on, people used hutches or chifforobes to stuff away their equipment. Before long, furniture manufacturers noticed that living rooms across North America were piling up with equipment and videotapes. They developed the big brother of the stereo cabinet, the entertainment center.

Entertainment centers are available almost everywhere from a myriad of companies in almost any kind of configuration and quality. You can find them at your local appliance store, discount department center, or furniture outlet. So, what's the point of having a section about them here? I know that some people will be put off by the "common sense" nature of this section, but thousands of poorly made entertainment centers are bought and later thrown out or dumped at second-hand stores. Overall, more money might be wasted on poorly made entertainment centers than any other single type of video equipment.

Like every piece of electronics equipment, check to see how long the warranty lasts and what it covers. A great warranty doesn't necessarily mean that the entertainment center is good, but it's a start. The best sign of good-quality equipment is if solid wood was used in its construction. Solid wood is best because it is generally stronger, more resistant to warpage, and can easily be refinished. And, in my opinion, it looks better. You can find real-wood entertainment centers at unfinished or regular furniture stores, or at some outlets where locally manufactured furniture is sold (craft stores, antique malls, etc.).

If the entertainment center isn't made with solid wood, chances are that the bulk of it consists of particle board (also known as *flakeboard* or *pressboard*). Particle board is a general name for some type of board that has been manufactured by grinding or tearing up small pieces of wood and pressing them into a solid board. This stuff is often used because it's much cheaper than solid wood. A particular grade (pardon the pun) of particle board, called

MDF, is specifically used for speaker cabinets because it is denser, heavier, and more uniform than either real wood, plywood, or the lesser grades of particle board.

Particle board is generally available in four different general types:

* *Wood veneer* A wood veneer is glued over particle board pieces.
* *Plastic laminated veneer* A veneer is manufactured with a simulated wood design impressed in the plastic. This veneer is glued over particle board pieces.
* *Paper veneer* Plastic-coated paper is printed with a simulated wood design. This paper is glued over the particle board pieces.
* *Painted* Sometimes particle board is covered with a heavy coat (or several coats) of paint, usually either black or white, with plastic-laminated edges.

Aside from the general comments about the style of the furniture, it is wracked with some other problems. The biggest problems with particle board furniture are:

* If the finish is cut or worn through, it is virtually impossible to fix.
* It is prone to severe warpage.
* It is heavy in weight.
* Boards can easily chip or break, and this damage is impossible to fix.
* If the unit is disassembled once or twice, the screw holes are prone to tearing out. This damage is difficult, if not impossible, to repair.

If you really think that a particle-board entertainment center is the way to go, carefully choose a system with better-quality materials—don't just buy something from a picture in a catalog. Wood- or plastic-laminated particle board is a much better selection because it will look better, be more resistant to finish wear, and hold the particle board together better. In the latter case, the laminates prevent the particle board from breaking or chipping, a valuable protection. The paper- and paint-covered furniture is the cheapest, beware! It doesn't look as good and it is the easiest to damage. In my opinion, these two types should be avoided at all costs.

FUNCTIONALITY

One of the dilemmas of buying an entertainment center is functionality. You want to purchase some well-made furniture that looks nice, but it must be large enough to handle all of your stuff and it must be functional so that you can file away and access everything easily. Sometimes, the solid-wood entertainment centers push form way ahead of functionality, so be sure to look for the features that you want (Figs. 7-1A and 7-1B). Some of the points to check include:

Fig. 7-1A A fairly compact home entertainment center. Sanyo Fisher

* The TV must be able to fit in or on the allocated cabinet space.
* The racks must be large enough to fit all of your A/V equipment.
* It should be able to hold an acceptable amount of your video collection.
* Your videos should be efficiently stored, yet easy to access.
* The doors and drawers should operate smoothly.
* Are there any other ingenious concepts here?

Fig. 7-1B A larger, but still fairly compact home entertainment center with space for equipment, speakers, and videos. Sanyo Fisher

AUDIO QUALITY

It might seem kind of silly to think that your entertainment center could have an impact on the audio quality of your home enterainment center. The entertainment center doesn't play a large role in audio quality, but it can have a significant effect in at least one area: stereo separation. In some entertainment centers, especially those from audio/video equipment manufacturers, the main speakers are mounted on either side of the TV. This significantly reduces the stereo separation and width of the sound stage. Instead of the sound hitting you from all sides, it hits you from one general direction. In defense of these entertainment centers, the manufacturers construct them for ultimate convenience

Fig. 7-2 The Heathkit 1967 version of an entertainment center: the entertainment wall. At the top is the picture tube, to the bottom left is a radio receiver/amplifier, and to the lower right is a panel containing all of the TV controls.
Heathkit

and compactness--if they built it any other way, people would complain that the system was "too messy" or "large."

If space is at a minimum and a good-sounding TV is OK with you, stick with this type of compact entertainment center (especially if it came along with your home theater package), but if you're trying to milk the system for everything you can get, follow the speaker-placement tips in Chapter 5.

IN-WALL MOUNTING

Just for the sake of creativity, an alternative to the traditional A/V cabinet is included here: the A/V wall. The inspiration for this is the 1967 Heathkit electronics catalog (Fig. 7-2). At the time, Heath's business of selling electronics kits was booming. Electronics hobbyists from around the world would buy, for example, a Heathkit radio, and would assemble it themselves to save a bit of money and primarily to learn more about electronics. For the advanced kitbuilder, Heath offered a color TV kit. The catalog offered several different TV configurations, including building everything into the wall. This concept seems a bit peculiar now (Fig. 7-3), let alone more than 30 years ago!

MORE INFORMATION

The best way to figure out what you want is simply to look around and see what's available. Just like searching for A/V equipment, go to a few different stores and take note of what is in stock. Look at the features, the materials that were used, and the way that everything is made, and try to figure out how it would fit into your home. To have a really quick look through at some entertainment centers that might give you some great ideas for what you want, see the following Internet Web sites:

William Laberge Cabinetmaker
http://www.vtweb.com.laberge/entertain.html

Woodtown Unfinished Furniture
http://www.woodtown.com

Fig. 7-3 The turn-of-the century version of the "entertainment wall." In this case, the equipment racks have been built into an area that is recessed into the wall. Pioneer

The Oak Country Peddler
http://www.amishfurniture.com/ocpnta.html

Diamond Case Designs
http://www.diamondcase.com/tt/tt.html

The Closet Factory
http://www.theclosetfactory.com/storage.htm

THE SOUND STAGE

The art of setting up a great home theater system in your house isn't just limited to organizing everything like a compulsive librarian. One very important aspect of the audio in your home theater that is almost entirely based on the speakers and their stereo separation is the sound stage. The sound stage is the positioning of the sound, as you hear it (Fig. 7-4). The positions that affect the sound stage are front and back (depth), up and down (height), and left and right (width). The sound stage is very important when designing and installing any home theater system because it is the key to making the audio from your system sound life-like.

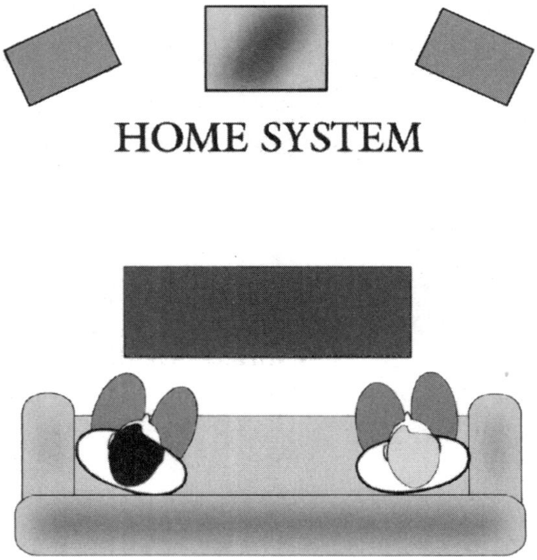

Fig. 7-4 The typical sound stage for a home audio or home theater system.
CD Technologies

The combination of the physical speaker locations, the stereo effects of the recordings that you listen to and the combination of audio signals in the air creates the sound stage, which is much different than the physical speaker locations alone. When you have two (or four, in a surround-sound system) widely separated speaker locations, the audio will not sound like it is coming from each of the sources. Instead, it will spread out and fill the room. If all of your speakers are mounted in the front of the room, the sound stage will be decidedly to the front, which is generally good; without rear speakers, the sound will not be as realistic because the audio will lack depth. As a result, it is best to have a balanced set of speakers around the room. With some creative speaker placement and acoustic design, you can create several different sound stages.

To create a proper sound stage, you should have the two speakers mounted near the same wall, both facing out, toward where you will be sitting to listen to the stereo. The speakers should both be separated by a considerable amount of space, which depends, in part, on how far away you will be sitting.

ROOM REFLECTIONS

One potentially serious problem for the listener is that of sound reflections. Sound will readily reflect from flat, hard surfaces. If you have two hard, flat surfaces that are parallel to each other, the sound waves will emanate from the speakers, bounce off one wall, reflect off the parallel wall, reflect back to the opposite wall, etc., until the sound dissipates. A worst-case condition occurs in a house: a typical room features three sets of parallel surfaces, all of which can wreak havoc on an audio signal (Fig. 7-5).

You have probably experienced this type of echo when you walked into a completely empty room (with a hard floor), basketball court, or racquetball court. If you talk in a typical empty room in a house, the echo is very fast, making your voice sound a bit fluttery. This echo should quickly dissipate. If you talk loudly on a typical indoor basketball court, the echo will be long and strong; you should be able to shout "Bach" and hear at least one repeat of the word before it dissipates. The college that I went to contained one poor, much-oversized, all-concrete racquetball court. This thing was like a terrible echo chamber. If you smacked a ball off a wall, the sound wouldn't dissipate for at least 10 seconds. If you wanted to talk to

Fig. 7-5 Not only does this room look great, but the room was designed to break up sound. The walls aren't parallel and the ceiling is vaulted, with exposed beams to further dissipate the sound waves. Pioneer

someone on the court and be understood, you had to wait for the other noise to dissipate, walk to within a few feet of them, and speak quietly.

From these examples, you might have picked up some of the elementary principles of room acoustics (if you didn't know them already):

1. Some building materials reflect sound waves more efficiently than others.
2. The size of the room affects the time delay of the echo.
3. These audio reverberations will distort a natural sound.

One positive aspect of this scenario is that you don't have to make the parallel surfaces nonparallel (massive remodeling!). Instead, you can experiment with interior decorating to dissipate sound. Carpeting on the floor automatically eliminates the parallel echoing surface between the floor and the ceiling (Fig. 7-6).

Fig. 7-6 Although the minimalist design of this home theater system might be attractive to some, the many flat, parallel surfaces will not dampen the many sound reflections. Even the carpet will be relatively ineffective because it is too small and thin. Solid Solutions

Now, you've got two more sets of parallel surfaces to conquer. A large bookcase will successfully dissipate much of the sound energy because there will no longer be one flat surface, but instead many jagged edges. Along the same lines would be a wall of slatted bifold doors or shutters. Some of the other absorptive possibilities that might fit in with the decor of your house are: a woven blanket or rug (Southwestern style), a quilt (Colonial or "country" style), or even a large bear skin (Adirondack style). Some high-end audio companies sell sound-diffusing materials for your house, but they are terribly expensive and, in my opinion, they are ugly. Some of the possibilities are sections of wooden grating for the walls and huge carpeted columns that are placed in corners to dissipate corner reflections. Using typical audiophile supplies, your listening room would look slightly more homey than the rooms from the set of "Battlestar Galactica."

Between the preceding solutions and the addition of a couch and a few padded chairs, you will probably find that the sound in the room is not noticeably degraded by room reflections. It might not sound as good as if you had covered the walls and ceiling with acoustic tile, but the solutions listed here are much more practical, nicer looking (by most people's tastes), and far less expensive.

Audio room tweaks are both an art and science, but the end result is that you need to move the speakers and sound-deadening objects around the room for the best results. If you would like to check more into the theories behind room acoustics as they relate to audiophile stereo systems, see *The Complete Guide to High-End Audio* by Robert Harley. For a look at the subject from the prospective of a professional recording engineer, see *The New Stereo Soundbook* by F. Alton Everest and Ron Streicher.

DARKNESS

For the audiophile, a perfect system must be absolutely quiet, with no system noise, and the listening environment must be equally noise-free. So, noise from the street (passing cars, people, etc.) should be eliminated, as well as noise from such things as furnaces, air conditioners, and other appliances. Not only is sound dampening covered here, but it is included in other references, such as magazine articles, scientific texts, and the books mentioned in the previous section.

Even if you only typically watch TV at night when light noise is virtually nonexistent, you still might have some occasions to duke it out with glare. For example, weekends and holidays are prime candidates for TV viewing. And you sure don't want to miss another "Dolly Parton Country Christmas" special, now do you?

One counterpart to sound dampening that I've never seen reported is light noise (light dampening?). Sound dampening is necessary to prevent audio distortion and outside noise that would compete with your audio. Although you shouldn't have to worry about problems from light reflections (unless you have mirrored walls in your house), light noise can be a serious problem. In the past, glass sheets were expensive, so windows were either used sparingly or were multi-pane units that were typically installed one or two per room. But houses from the past 20 or 30 years use much more glass—sometimes entire walls consist of glass and wood or steel structural supports. The theory for such design is simple—if you place your house in a location with a beautiful view over a valley or in the woods, why not enjoy it? Likewise, the rooms with the most glass are the ones that receive the most use, typically the living room. Of course, the living room is also the place where the home theater system is often placed (Fig. 7-7).

Such a great amount of light will compete with the strength of the TV video. Unless you have always had a dark location and never watched TV at someone else's house, you should know exactly what sunlight "noise" looks like. Bright sunlight on the screen makes the picture look light brown, with a little bit of color and movement here and there.

One improvement that TV manufacturers are developing is the black or "extra dark" screen. These darker screens make the picture look brighter, whether in daylight or in bright lights. Of

Fig. 7-7 Sunlight glare would be the least of your problems in this windowless home theater system. I can't be sure where this one is located, but my bets are on the basement. THX

course, although this technology improves the picture, it is no cure for direct sunlight. The only cure for direct sunlight is to physically eliminate it.

CURTAINS

The only temporary way to eliminate sunlight is to install curtains in front of the offending windows (Fig. 7-8). This can be a serious problem because curtains could ruin the decor of a great living room. Also, the curtains must be lined or the sunlight will still blaze through. The best curtains for the job are either the super thick canvas ones or those that are lined with black velvet. Neither type of curtain is terribly common because they are very expensive to make and are out of style. Another set of problems with this type of curtain is that they are both heavy and bulky. Not only will such heavy curtains be very noticeable—even when closed, but they also require heavy-duty fasteners to hold them up.

If you have a bad lighting problem in your home theater room,

Fig. 7-8 No windows are opposite the TV in this room, so the biggest potential problems would be sun in your eyes at dusk, rather than screen glare. Pioneer

check with some of the home theater distributors in the Appendix. Some of these companies actually sell curtains that were designed specifically for home theater applications. And you thought that home theater only meant a big TV connected to a hi-fi audio system!

TINTED WINDOWS

Another possibility to reduce light glare is to have tinted windows throughout your house. Installing these windows might seem like a great idea if you are currently building your home or are about to do so, but it is a terribly expensive remodeling change—especially just to reduce the glare on your home theater system. A better solution would be to try moving the system so that the TV would be in a location with less glare.

Still another option is window tinting. Some companies are now offering do-it-yourself window-tinting kits manufactured by 3M. I believe that these are all tinted translucent sheets of adhesive plastic that you can attach to your windows. For most people, the big advantage of using this stuff would be heating efficiency: less sunlight would heat the house in the summer, and the plastic would prevent some of the heat from escaping through the windowpane in the winter. Of course, for the home video enthusiast, the reduced glare would also sweeten the deal. I haven't tried any of these kits yet, so I can't recommend or condemn their use. For more information, contact your local hardware store or see the Wow Enterprises Internet Web page for their supply of glass-tinting services:

http://users.nova1.net/~wow

MOVE

The last possibility for the glare-conscious home theater junkie would be to move the system to somewhere else in the house. Unless you have plenty of empty rooms in your house, this might not be a possibility. Still, if you have an unfinished basement that you've been eyeing up, this might be the nudge that you need to build a new system in your basement.

Our last house was built around 1910 and the basement consisted of a slopped-in concrete floor, stone walls, and rafters at about 5 feet 8 inches. Finally, the place was filled with a monstrosity of a furnace. I wouldn't entertain anyone in a place like that! So, first you've got to have a good basement to work with before you can consider sticking an entertainment center in it. If you have a dry basement with a smooth concrete floor and tolerably high ceilings, you have the makings of a great basement video room.

Basements are choice locations because:

* *Low noise* With walls surrounded by dirt, this should be the lowest-noise location in any house.
* *Low light* Basements usually have either no windows or tiny pull-out windows, so they are dark most of the time—perfect for home theater.
* *Climate* Because the basement walls are insulated with a huge amount of dirt, very little heating or air conditioning is required to keep the climate comfortable.

The worst problem with using basements for living space is that nasty combination of leaking water and humidity. If your basement walls leak, try to fix them by using basement wall sealant. If they still leak or bead water, you might be able to solve the problem by digging a footer out along the outside wall where the leak is, then tarring the outside of the wall. Finally, you can fill some of that footer trench with gravel to keep the water away from the wall. Or you might decide that it's best to go with a licensed contractor to seal up your basement.

To keep dampness from seeping up through the floor, coat the whole floor with tar, place in 2 X 4s or 1 X 2s as "joists," and lay in the subflooring overtop of that. For more information on construction and basement remodeling, see *Housebuilding: A Do-It-Yourself Guide* by R.J. DeChristoforo, *Carpentry & Construction* by Rex Miller and Glenn E. Baker, and *Converting Basements, Garages & Attics* by R. Dodge Woodson.

PUTTING TOGETHER A SYSTEM

So far, this book has covered many different home theater topics, but here's a look at a sample system, just to show what is involved in putting a system together. This is not necessarily an optimum system for all situations (there is no such thing), but it will give you some ideas about what you should be doing (or not doing) when you design and build your own home entertainment system.

THE EQUIPMENT

This system is designed around a 32-inch Samsung TSF3278 TV

with picture in a picture and S-VHS input. To provide video playback, a Pioneer DVL-500 DVD player and a VR730HF S-VHS hi-fi VCR were chosen. The loudspeakers are the Rock Solid Sounds HCM-2, and the A/V receiver/amplifier is the Harman/Kardon AVR80. This system is a bit expensive, costing several thousand dollars, but none of the components are highly priced, compared to the rest of the market. And, with a little compromising on lower-priced components, searching for used units, or sales and closeouts, you could build a comparable system for considerably less money.

A 32-inch TV isn't as large as many of the units on the market (and it's not nearly as large as a projected image), but that's still a nice, large area, without filling a whole wall. This particular TV will play back up to 700 lines of horizontal resolution, so although it isn't up to the 1000-line HDTV standard, it can easily handle the higher-resolution output from a DBS TV broadcast, a DVD, or an S-VHS tape. In short, this is everything that you need to really roll the film!

INTERCONNECTIONS

Shielded RCA cables with screw-on plated metal covers are then used for all of the audio runs. Then, the equipment is connected as follows:

* The S-VHS output of the A/V receiver goes to the TV S-VHS video input.
* The outputs of the DVD and the VCR each go to the inputs of the A/V receiver.
* The outputs from the TV go to the A/V receiver.
* Everything is plugged into a surge protector power strip that is insured for at least as much money as the equipment is worth.
* All of the equipment is properly grounded to an outside ground (see Chapter 6).
* One output from the A/V receiver goes to the VCR input so that you can record programs (Figs. 7-9 and 7-10).

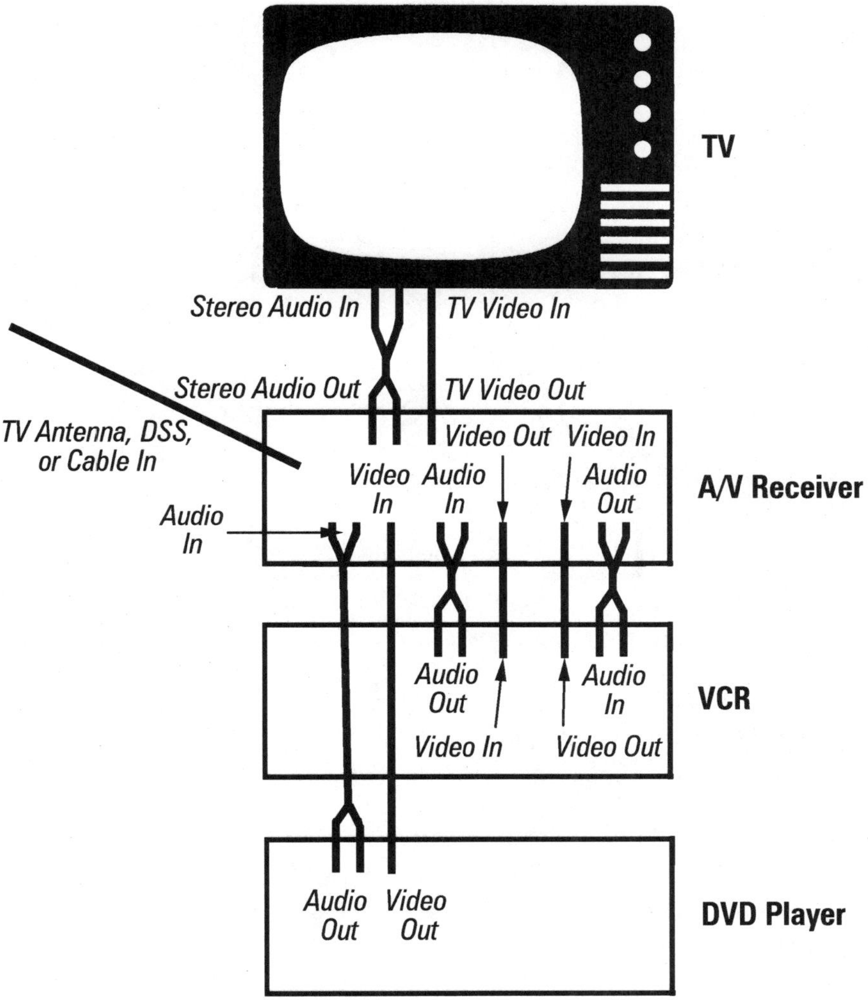

Fig. 7-9 The audio/video input/output cabling diagram for a TV, A/V receiver, VCR, and DVD player.

FURNITURE AND LAYOUT

After getting this much equipment and connecting it together, it's only a matter of having it in the right place. For this system, we're starting with a completely new basement room. Why not, it's hypothetical anyway! This room is started fresh from the ground up: new floors, new walls, and a new ceiling. Because this is an all-new system, the cabinet areas are being built right into the wall, rather

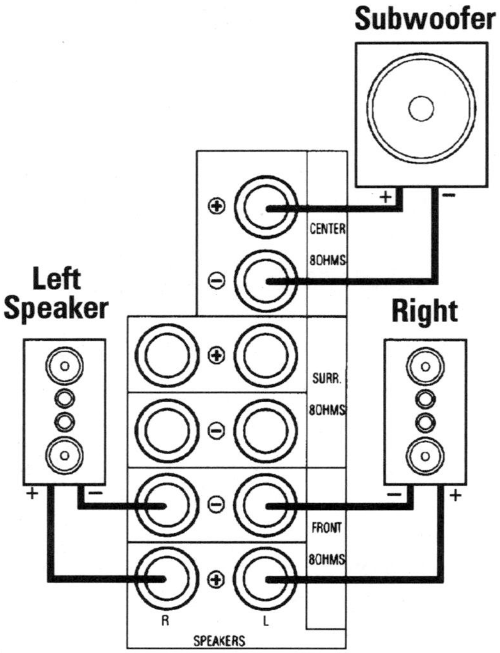

Fig. 7-10 The speaker output connections for a system with a single pair of stereo speakers and a subwoofer. Harman/Kardon

than spending loads of money on an entertainment center. Also, this method saves more space and the wires can be run in the space behind, rather being looped around shelves or being pushed through holes in the entertainment center.

All of the inside walls are built against the cinderblock walls of the basement, except the wall that will contain the A/V equipment, which is built several feet out from basement wall. The entire wall is constructed out of cabinets, drawers, and fold-outs. This system can be either home-built, constructed with entertainment center or kitchen cabinets, or use a home-built structure with drawers and doors from kitchen cabinets. The whole dilemma is finding cabinets that you like from someone who is willing to sell them door by door!

There aren't too many surprises with this entertainment center. The center top cabinet contains the television set. The inside of the cabinet was made 38 inches wide so that a 35-inch television could be purchased to replace the 32-inch model sometime in the future, if necessary.

The cabinet to the left side of the TV cabinet is where the equipment will be held. To properly ventilate the equipment, this cabinet has no back, just a single crossbar so that none of the components can get pushed out the back and onto the concrete floor behind! Because the space behind the room will be unheated (at least not directly), it should stay between about 50 and 65 degrees. This will keep the equipment nice and cool, yet you won't see the space behind because the raised-panel doors will hide it. If you don't like the space behind it wide open, you could drill a bunch of holes in a piece of plywood and tack it in place behind the cabinet. To save space and offer more flexibility, this cabinet contains no shelves for the equipment; the pieces can be stacked on top of each other.

Under the TV cabinet is the video library. Here, 12 sliding VHS video racks are installed against the side walls. If the cabinets are 22 inches deep, the video racks could hold as many as 240 videocassettes. If necessary, more racks could be installed in the cabinets on either side. They are left open in case the video library drastically increases in the future.

Finally, the last cabinet area to the far left is basically open. It has a mini cabinet on the top and a little desk surface. For more privacy and a better view of the TV, try making a door for the front that will drop down and become the table surface. In this little deck area, you can either place a home computer system or (better yet), pop up a laptop and try some on-line video editing. With a hole drilled into either the side wall or the back, you can easily interconnect the computer with the other pieces of A/V equipment. Or you could add a video editor and another VCR to semi-professionally edit movies or video segments.

Everything is set, except for the cable that runs to the speakers. Because this theoretical system is being designed from the ground up, we have the flexibility of running cabling before the walls have been finished. As mentioned previously, this is a great advantage! In this room, a two-inch-diameter PVC pipe is run inside the ceiling, between the rafters. The pipe's entrance is at the edge of the entertainment center wall. This way, you can go to the space behind the wall and have access to this end of the cable run at any time.

Another PVC pipe can be run down the inside of the wall that is opposite from the home theater wall. About 18 inches up from the floor, this pipe should end into a plastic electrical switch box. Four lengths of insulated, shielded two-conductor speaker cable are run though the pipe, from the entertainment center to the switch box. The wall behind the entertainment center will be left open to the walking space behind it, so there is no need to run the wires through a pipe in this section. At the other end, the cables are all run through the switch box, leaving a few extra inches of overhang. The wires are stripped for about an inch at the ends. Then, holes are drilled through the blank switch cover, good-quality jacks are installed in the switch cover, and the wires are screwed in tightly. Because only two speakers will be used in the rear positions, two of these sets of wires and jacks will be unnecessary. But you never know when an extra audio or video cable might come in handy...and it's much easier to run it now, before the room has been finished!

Then, the rest of the room construction is finished: drywall is nailed in, the wall compound is slopped and sanded, the walls are painted, the ceiling and carpet are laid in, and the cabinets are finished. After the room is ready to go, you can place the speakers in the positions that you want them in. Now, you can measure the length of the cable from the speaker positions to the jacks that are mounted in the switch cover. Be sure to leave some extra length in the cable so that it can be run along the edge of the wall and be out of the way. Then, the plugs can be added, as described in the last paragraph.

CONCLUSION

Although having the equipment for a home theater is the basic core of the system that you will have, the way that you arrange it will have a major impact on how you use it. With everything set up, you only have to worry about keeping everything in proper working condition. See Chapter 12 for more information on maintenance and simple repair.

8 COMPUTER VIDEO

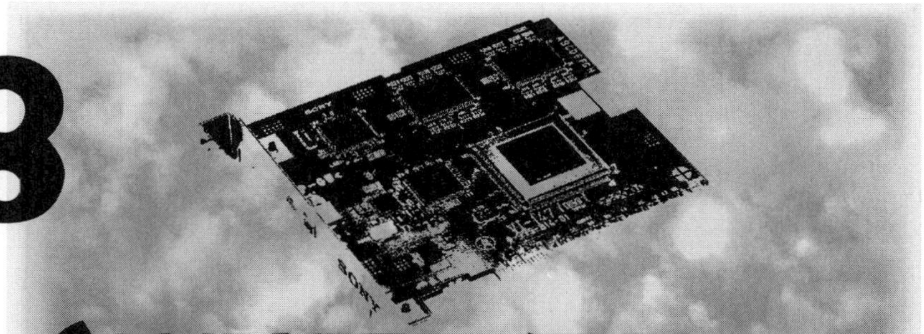

For years, getting that home video experience meant sitting in front of the TV. But, in the 1980s, the typical stereo system had morphed into something different—the home theater. Here, you could sit back, catch an album, and then catch a movie, complete with hi-fi sound. In the 1990s, new video technologies, such as the laserdisc, DVD, DBS broadcasting, and high-resolution wide-screen TVs dramatically improved the quality of video.

An entirely new listening system has been developed in the 1990s—one that has developed separately from the video sources of the past. If you read the title of this chapter, you know it is the computer video system.

Considering the state of the home computer in 1985, it is just amazing that computerized video systems became possible in less than 10 years. At that time, most home computers were running at a fraction of the speed of today's machines. The most common computer storage devices were cassette players and low-density single-sided 5 1/4" floppy disks. Even by the late 1980s, it was just exciting to be able to play Dig Dug or Pac Man on a computer and hear a really tinny version (sounding like today's musical greeting cards) of the video-game theme songs (Fig. 8-1).

Computers are becoming the next big thing in the video world because of the development of smaller integrated circuits and higher-capacity hard drives. The increase has been blistering: nearly every year the processor speeds are doubling. At this writing in late

1997, it's 300-MHz MMX technology, in 1996, it was the 200-MHz Pentium; and at the end of 1995, it was the 133-MHz Pentium. In addition, the price cuts are staggering. That Pentium 200 computer might have cost $4500 when it hit the market, but it's now well below $1500. The 486 DX4-100 computer (roughly half the processor speed of a Pentium 100) might have initially sold at the same price as the Pentium 200, but now you can find it for $500 or less.

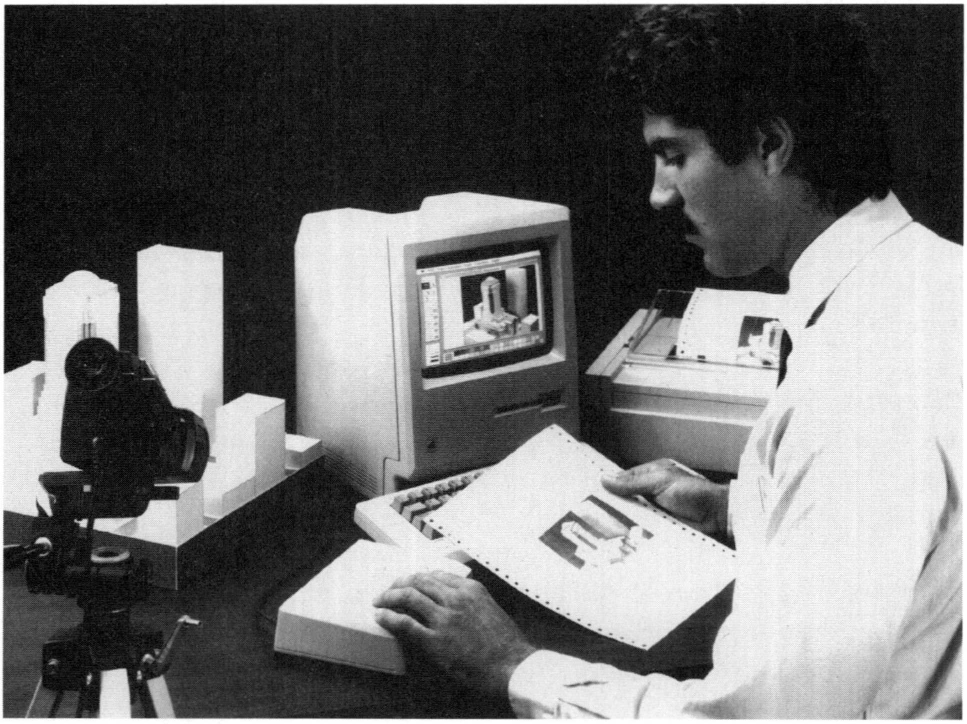

Fig. 8-1 This computer "video" (still images only) might have seemed pretty hot 10 years ago, but now it's hard to even imagine this equipment in use, let alone for video applications.

The same kinds of increases in processor speeds and price cuts are occurring in the rest of the computer industry. CD-ROM drives, sound cards, RAM chips, software, etc., are all becoming faster and less expensive. Multimedia, the catch word of the 1990s, is no longer used to impress your friends. It's so common that everyone just expects multimedia from a computer.

Now that multimedia has become mundane (but not an "old" technology) and computers are filling homes and offices around the world, the emphasis is not on the novelty of multimedia, but whether a computer can be at the heart of a video system. The answer is, of course, yes (with a few reservations).

The type of system described in this chapter is a digital/analog combination with a multimedia computer, a DVD drive, an amplifier, speakers, and a VCR. Everything is centered around the computer (in terms of use and applications) and the computer system (in terms of location), which is why it is considered as being separate from a "traditional" home theater system.

A standard multimedia computer audio system would consist of a computer with a CD-ROM drive, soundcard, and amplified speakers. Such a configuration will work for computing, but not for video playback. The focus of this chapter is to add video to the computer and dramatically improve a system that most people take for granted.

HARDWARE

The following computer and audio hardware is either necessary or helpful when constructing or improving a computerized DVD-ROM video system. Remember that the requirements for running full-motion video and CD-quality audio are much higher than those of the old multimedia MPC Level 2 specifications.

MOTHERBOARD PROCESSOR

In the early to mid-1990s, the speed of a computer often prevented users from successfully installing and using multimedia upgrade kits. Many people were concerned about the amount of RAM required and the minimum processor speed. If you have purchased a computer that was manufactured since 1996, chances are that the computer will have no problem with the processor speed. The minimum requirement is a motherboard with a 100-MHz Pentium processor.

HARD DRIVE

If you are just playing DVDs on your DVD-ROM drive, then the amount of available space on your hard drive is basically irrelevant. You only need a minimum of about eight megs of available hard drive space to be able to set up and run the DVD drive. That situation only changes if you are saving video as computer files (typically .AVI files) or mixing/editing audio or video. You just don't need a little bit of empty space, you need a massive hard drive! A 30-second chunk of sound, recorded on your computer in stereo at CD quality (44.1-kHz sampling rate) will require 5.25 megabytes of space on your hard drive! Imagine how much would be necessary to record full-motion color video as well as stereo audio! Lucky for us, the cost per megabyte of hard-drive space has dropped incredibly over the past few years (Fig. 8-2).

Fig. 8-2 The key to modern computer storage, the hard drive (shown with the case open). Maxtor

Even though a 160-megabyte hard drive is required for a computer to pass the MPC Level 2 specification, few new computers are sold with less than 10 times that much available space! As of this writing, most new computers are being sold with disk drives that range in capacity from 2 gigabytes (2000 kilobytes) to 6.4 gigabytes (6400 kilobytes). That might sound like a tremendous amount of available space, but if you perform much multimedia work (audio, still images, and video storage and recording), that space will be filled up before you know it.

SOUND CARDS

Sound cards are the heart of the computer sound system. They act as a digital-to-analog (D/A) converter, analog-to-digital (A/D) converter, MIDI interface, and a central "station" for audio interconnections: line in, mic in, audio out, and MIDI in/out. In short, they make it possible to output or input audio.

Current models of sound cards all use wavetable synthesis, as opposed to some other methods of deriving sound, which had been used in the past. At this point, the only thing that you need to know is that wavetable sound cards will produce realistic audio, and some of the other now-defunct types (such as FM synthesis) produced video-game-type sound. FM synthesis sound cards might be nice for playing "Asteroids," but they aren't much good for playing back CD-quality audio.

The D/A converter part of the sound card works on the principles that were covered in Chapter 3, which covered digital audio from digital discs. For more information on D/A and A/D conversion, see Chapter 3. To be completely superficial and nontechnical, let's just say that the D/A converter changes all of that binary code into great music.

SPECIFICATIONS

Of course, when you are considering building an audio system around your computer, you have to be concerned with the quality of the sound output. You won't be able to go into a computer store and listen to every model of sound card with a great amplifier and pair of speakers, so you'll have to rely on the specifications (and possibly a review or two) if you want to be satisfied.

Sound cards are intended for different computer applications. Some focus on a bajillion different instrument sounds for MIDI. Others provide neat gimmicks that you can use to manipulate audio (such as reverb, chorus, and specialized audio software). Still others simply strive to be inexpensive. But a few work hard at providing high-quality audio output. You can see this in the specifications.

One company that has consistently worked toward producing audiophile-quality sound cards is Turtle Beach. For example, their MultiSound Pinnacle sound card features the following specs:

* S/N: > -97 dB
* THD: < 0.005%
* Stereo crosstalk: 90 dB at 1 kHz
* Frequency response: 10 Hz to 22 kHz

As you can see from the specifications, you can assume that this sound card has a better sound than many of the CD players currently on the market. In addition, the MultiSound Pinnacle can also record as high as 20-bit audio with a frequency of 48 kHz. This higher-resolution audio is even better than that which is available via CDs (16 bit at 44.1 kHz)!

VIDEO GRAPHICS CARDS

Sophisticated video graphics cards are essential for playing back video or computer graphics from regular programs to games or especially standard video. These cards create all of the parameters for video output: the number of colors used, the resolution of the screen, etc. With memory installed right on the board, it is able to store video while data is being pumped out from the DVD, CD-ROM, or hard drive. The minimum requirements for DVD video are the same as those for the SVGA specification (except with more RAM): 256 colors, a screen detail of 640 x 480, and 2 megs of RAM. But, this is no problem for most video cards on the market. They can now display as many as 16 million colors and have as much as 8 megs of RAM. Other proprietary technologies are used to improve the graphics performance, as well. Because the video graphics card is the interface between the hardware and the screen, it is one of the most important components of a computer video system.

MULTIMEDIA ACCELERATOR BOARDS

Unlike the other components featured in this section, multimedia accelerator boards aren't used for a specific application. It's just like the BASF ads : "We don't make the snowboards. We make them faster." Multimedia accelerator boards do just that; they make multimedia play back faster. In a sense, the boards act as a computer in a computer--the extra computing power is used to enable the unit to process more commands. One manufacturer of these boards, Mpact, seems to emphasize using multimedia accelerator boards for use in computers with DVD drives (Fig. 8-3).

Fig. 8-3 The Mpact Mediaware hardware/software multimedia accelerator combination. Mpact

MONITORS

Like the standard home theater system, the actual output device (in this case, a monitor) is important because this is where you watch the video. But there's a bit of difference between TVs and computer monitors. The important difference is interlacing. As mentioned in Chapter 2, conventional TVs are currently interlaced (every other line is shown blank). Computer video standards require noninterlaced video (the lines shown are always conveying video information). This difference means that the two types of video are not compatible, and also that computer video output is of better visual quality.

Because the quality is so much better, you can comfortably sit close (within five feet) of a computer monitor to watch video. This makes viewing video at a computer more of a personal than a group experience, which can be either good or bad. The good point is that you don't need much space to enjoy a good movie—just a desktop or laptop computer and a place to sit. The downside is that a bunch of people can't watch a movie at one time. This problem is confounded by the cost of computer monitors, which typically only range in size from about 14 inches to about 21 inches. Even at only 21 inches, a computer monitor will often cost well over $1000—more than most 32-inch TVs!

Unlike the situation with TVs, you don't often need to spend much time checking computer monitors to be sure that the picture is sharp and the color looks good. The requirements for SVGA computer video quality are superb—much better than those for NTSC video.

The final analysis for the monitor's quality can often be determined just by reading the statistics for the unit. Aside from the size of the screen, the resolution, and the price, you should also look at the dot pitch and the viewable area. The *dot pitch* is the distance between each dot on the screen. Typical spacings range from 0.25 milliameters to 0.28 milliameters. The screens with the smaller dot pitches will look most realistic, and the others will look more "TV like." The *viewable area* is the size of the screen that video is actually displayed on. For example, most monitors have a viewable area that is about one inch smaller than the published size (e.g., a 17-inch monitor might have a viewable area of 16 inches or 15.7 inches).

If you are choosing a monitor for your computer, the big concern will be the monitor size. Should you buy a DVD drive and a big monitor for your computer and emphasize that as your main movie-watching area? Should you go with a regular-sized monitor (15 inches) and a DVD drive in your computer so that you can catch movies there when necessary and a DVD deck for your home theater system? Or should you get the DVD drive and integrate your computer system with your home theater?

Some monitor manufacturers, such as Princeton Graphics, have worked steadily at integrating television/video monitors with computer monitors. Some of these units operate as both full-screen televisions and as computer monitors (Fig. 8-4). These models are especially handy for families who need to consolidate their resources, those who use Web-TV, or for business people who need their monitors for computer or video applications.

Fig. 8-4 Computer games and TV at the same time? A dream come true for the tough-to-distract crowd. Princeton Graphic Systems

But this isn't the only way that computer monitors are evolving. In addition to this sort of convergence between computers and television, new types of video displays are being specially formulated for the virtual reality and computer game markets (Fig. 8-5). These

Fig. 8-5 A specialized monitor for computer-game applications. Samsung

displays (everything from VR goggles to LCD panels) will revolutionize the video of the next decade or so--even if they currently look like technological curiosities from a sci-fi movie (Fig. 8-6).

COMPUTER-TO-TV INTERFACES

If you plan to integrate your computer system and your home theater system, you'll need to buy a computer-to-TV interface so that your computer video can be displayed on a TV. This card is used to convert the computer signal into an interlaced TV signal and to compensate for the differences in resolution.

One of the best-featured computer-to-TV interfaces is TView Gold from Focus Enhancements. This card is expensive, listing at $595, but the whole screen is viewable (even at 1024 x 768 resolution) and it is compatible with NTSC, PAL, and SCART broadcast TV systems. You can see more information about their products at:

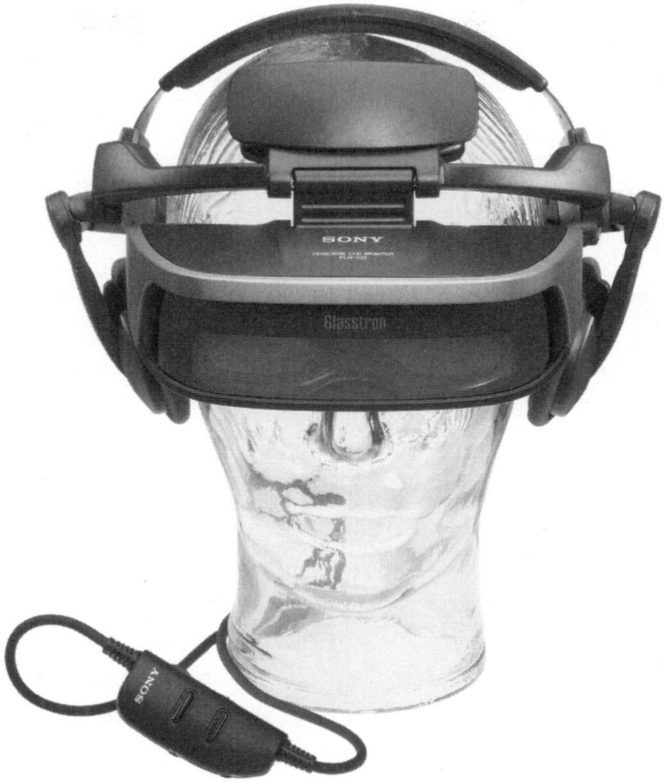

Fig. 8-6 The future of personal video viewing? The PLM-100 video goggles from Sony. Sony

http://www.focusinfo.com/home.htm

Two less-expensive alternatives are AVerKey3 Plus and AVerKey5 Plus from AVerMedia. These allow some neat functions, including zoom, pan, and scan freeze. They even include a remote control for computer chair potatoes. See more information about their TV and scan converters at:

http://www.allproducts.com.sup006/avermedia

Shop around for computer-to-TV interfaces because this technology is becoming increasingly popular, so more will be available and the prices will steadily drop. Some video graphics cards also have the computer-to-TV interface built in and a jack to output the signal to a TV. If you are buying a new computer system, it should be cheaper just to buy a good video graphics card with NTSC output than to purchase a video graphics card and a separate computer-to-TV interface.

AMPLIFIERS AND SPEAKERS

Amplifiers and speakers are some of the biggest rip-offs in the computer industry. The quality is generally bad, their specifications are often incorrect, and (except for surplus speakers) the prices are incredibly high.

The typical computer multimedia on the market today includes a pair of amplified speakers. The plastic boxes each contain one speaker that is about 3 inches in diameter and a small amplifier. Some of the better systems will also toss in an amplified subwoofer. Such systems typically cost anywhere from about $25 to $600. These systems either sound so bad that you wouldn't want to use them or are so expensive that you couldn't afford to.

The bottom line is that multimedia speakers are inferior by design and the prices are inflated. Plastic enclosures are never used in home audio applications, and amplifiers are never built into speaker boxes. A 4" "subwoofer" built in an 8" plastic box? Give me a break. These are extreme no-nos in home audio, and they are just as serious flaws in computer multimedia.

Fortunately, the problem is easy to solve. Just avoid specialized computer multimedia speakers and amplifiers. Instead, dig into the junk box. Your best bet on a cheap (or free) system that's better than most of the computer speakers on the market is an old junker music "system." That's right, the old 1980-era phono/radio/cassette system with the crappy speakers will knock the socks off of most multimedia speaker/amplifier combinations. The amplifiers were built to handle about 15 to 25 watts of power with somewhat low distortion—certainly not a Linn LK100, but substantially less expensive. Although the speakers are substandard, they are larger than multimedia speakers (not a true sign of better fidelity, I realize), and they have flakeboard cabinets with a larger internal volume. A better alternative would be to find a good 1970s or early 1980s era receiver for $20 to $50 and add some cheap, used bookshelf speakers. You could really beef up your computer's sound for about $100.

VIDEO CAMERAS

These days, it seems that most people who own a computer also own a VCR and a camcorder. It would be easy to just connect the camcorder to the computer and let'er rip. Unfortunately, it's not that easy. Nearly every camcorder available uses standard analog playback and recording. As a result, any video clip must be played back through the camcorder and converted into a video segment before it can be altered or saved on the computer.

Another strike against standard camcorders is all of the features (and their associated costs). If you only need a piece of equipment that can "take" a moving image (such as for videophone calls or videoconferencing), why would you buy the unnecessary VCR,

Fig. 8-7 Does videoconferencing sound like much more fun than just chatting on the phone? Here's a key piece of hardware. Sony

display, and special effects that come built in? Instead, you could simply purchase a digital video camera: a box with a lens and optical electronics. Of course, these units can't do much, but they are perfect for the simple videoconferencing applications for which most people are using computer video (Fig. 8-7). Presently, these digital video cameras start at $99 and go up from there (depending on the output resolution), but these prices are destined to drop significantly in the next few years.

Some of the latest still digital cameras now contain a disk drive so that you can save your photos to the disk, then simply pop the disk into your computer's floppy drive for access. Perhaps the future of video cameras is a digital video camera that would save everything to a rewritable DVD (DVD-RW) disc. Such a camcorder could be made extremely light, the disc would be rewritable and could store plenty of video, and (best of all for the video-editing enthusiast) everything would be immediately compatible for digital video editing on the computer.

VIDEO-CAPTURE AND RADIO/TV CARDS

Miniaturized computers, dedicated to a single application (or sometimes several applications), are used for all sorts of appliances—everything from microwaves to radios and TVs. Considering the plummeting cost of computers, sometimes it is less expensive to develop all of the hardware on a plug-in board for a computer, write software for the application, and let the personal computer do the rest. On the other hand, high-tech gadgetry pulls in lots of money, so some of the computer boards on the market are more along the lines of expensive novelties than something practical. Regardless, this section is about some of the communications boards that can be popped into your computer.

VIDEO-CAPTURE CARDS

Video-capture cards don't really have much to do with home video. They are used to "capture" video images from a TV (Fig. 8-8). This equipment snags the images one by one, as still pictures not as video clips. So, if this book was called *Home Photography*, then video-capture cards would have a place. But this little segment on video-

Fig. 8-8 The Sony DV-BK2000 board for video-capture applications. Sony

capture cards is included here because most computer TV cards double as video-capture cards. Not only can it be confusing at times, but both video cards and TV cards are often all lumped together in computer catalogs.

RADIO/TV CARDS

Radio/TV cards are the flipside of computer-to-TV interfaces. Instead of turning your TV into a computer, with one of these cards, you can turn your computer into a TV. Although it's a novel idea, it seems kind of like a sort of de-evolution—turning your "smart" computer into a "dumb" appliance.

Regardless of the Darwinesque technological implications here, the concept is kind of neat. The board doesn't simply turn your computer into a TV, it allows you to make the TV "screen" small and you can watch TV while performing different computing applications. Of course, the ability to watch TV on your computer would be considered a devious activity by many bosses and corporations, who would envision hundreds of employees snacking around while watching "All My Children." But for those of us who work at home, it's a convenient time-saver to be able to pop the screen up and catch some sports or news highlights while working.

These boards are typical computer-application boards: the board plugs into one of the open slots in your computer and the software does the rest. The boards contain a built-in TV tuner and a plug so that you can connect either an antenna or your cable TV service (Fig. 8-9). Aitech International manufactures several of these

Fig. 8-9 Watching TV on a computer. Productivity might be down, but at least the Pacers are winning! Princeton Graphic Systems

boards. For more information, see:

http://www.quikpage.com/a/aitech

For more information about computer radio boards, see the companion volume to this book, *Home Audio*, which includes a section about typical radio boards and even computerized communications/scanning receivers.

AUDIO AND VIDEO SOFTWARE

To do anything with a computer, you must have software (computer programs). Software is the interface between you and the computer;

via the software, you tell the computer what to do and (if the software is well written) the computer will tell you what it is doing and what your options are.

Every different computer application requires software, which only consists of codes on a disk. Thus, the process of developing new software is simple—the engineers must only create the code for the application. Unlike other manufacturing fields, they don't need to have a prototype constructed, have molds and dies cut, and other equipment set up. And, if the computer engineers want to add or alter some features, they just need to write the code and use the new information as the master for copying. If an engineer in car manufacturing would implement such a change, several of the machines would need to be altered to produce the new parts and the production protocol would also need to be changed. Because of these differences in processes, computer programming can be developed at little cost and updated regularly—this means that many different programs are available for nearly any application.

Computer video-related (or A/V related) software can be broken down into the following general categories: Internet video players, video players, and video editors. All of these jobs are necessary if you work extensively with video, but all of this work is too much for one piece of software to handle. Some of this software can be downloaded as "freeware," "shareware," or as a commercial software package.

Freeware Software that is absolutely free to use.

Shareware Software that is free to download, but a payment is requested by the author. Shareware programs are generally low-cost alternatives to expensive commercial software. Although the honor system is used for payment, some of these programs are either not fully functional or will stop working if the payment is not made in a given amount of time.

Commercial software Commercial software can be purchased in a package, complete with manuals and other information. These packages typically offer tons of features, and are usually compatible with other programs.

Internet video players are used to play back the different high-compression video and audio files that are placed on Web pages. The most common Internet video player is RealVideo, which is also known as RealAudio. The RealAudio media player will play back RealAudio—either in a live feed or a recorded segment. But, if the feed or recorded segment is RealVideo, the RealAudio player will play the audio and video. Unlike .WAV and .AU files, which require you to download audio to your hard drive, Internet audio players allow you to listen to programming in real time. Hundreds of radio stations are now live simulcasting their programming via the Internet, and now Internet video broadcasting is beginning to take form. One of the early trials for RealVideo was a live RealVideo-simulcasted show from Aerosmith. Aside from playback, Internet video players contain no features.

You can download a sample program of RealVideo/RealAudio from:

http://www.real.com/index.html

Standard computer video players will typically allow you to play a number of different formats of video, audio, and/or MIDI files. You can either create video files from a video-editing program or download them from either a Web site or from a CD-ROM. These video segments are handy for a variety of reasons: you can see anything from little entertaining film clips to videos from your family. Most audio players include some novelty features, such as looped playback (the file is played over and over in a loop), alarm (using audio files to wake you up at a predetermined time), or spectrum analyzer. Still, the most important feature of computer audio/video players is the ability to play back as many different formats as possible.

A few of the possibilities for playing back RealVideo are the shareware computer programs MM Player and MIDI Plus. Both of the programs are combination audio/video players. MM Player can play back .MID, .WAV, .AVI, and .MOV files. You can download it from:

http://home2.swipnet.se/~w-25992

MIDI Plus is freeware! It can play back .MID, .WAV, and .AVI files. You can download it from:

http://www.hivolos.com.cy/midiplus

The final category is that of video editors. The category of computer video players is somewhat small because many people like having the ability to edit and record audio and video, as well as play it back. Video recorders are fantastic tools because they give you the ability to edit as precisely as if you were using an expensive video-editing suite, yet they are much faster, less permanent, and allow you to listen to your work before permanently editing it (the mighty Undo command!). Video editors vary fantastically in their number of features and general quality. They vary from little shareware clip'n'paste programs to full-blown multichannel professional mixing/editing programs that cost into the thousands of dollars.

If a video editor seems to be in your future, check your priorities. If you plan to record professional video segments or programs for your clients, you would want to have multitrack options, many different audio and video effects that you could manually adjust, speed control, freeze-frame, animated graphics editing, and more. If you are recording music by yourself or if you are recording a band that uses synthesizers, then MIDI compatibility is a must, too. If you just want something that you can use to record some little video clips to send to your distant relatives, try a simple shareware editor that you can use to record video straight from a camcorder. Also, the ability to play back many different types of audio and video files will be helpful because you will probably use the editor to edit or play back some of the files that you find and download from the Internet.

The most popular commercial video-editing program is Adobe Premiere. This software can literally do almost everything: it can combine video, audio, animation, still images, and graphics to create movies on desktop for output to videotape, multimedia, and the Web. Premiere has enough special effects to do almost anything, and it runs seamlessly with other Adobe products, such as Photoshop, PageMaker, PageMill, and Illustrator. If you are a professional and use other Adobe products, Premiere is the software to purchase. Of course, the best comes with a price, more than $400 in this case. So, Adobe Premiere might not be for everyone because of the high price, but it is the prime video software for anyone who earns a living editing videos. You can download a demo of Premiere from:

http://www.adobe.com/prodindex/premiere/demoreg.html

If you are serious about video and are already using Adobe Premiere, you might want to spruce up your productions with some more effects. Although not true video-editing software, Adobe After Effects provides all sorts of special effects for users of the Premier package. Again, the price of perfection isn't cheap: about $700.

Videonics MediaMotion is a simpler program, but it has some neat features. One of the handiest is that you can use it to control VCRs and camcorders through the remote-control port for QuickTime digital editing. For more information about the software, see:

http://www.videonics.com

Another simpler program for the digital video hobbyist or professional is Asymetrix Digital Video Producer. With this software, you can edit two different video tracks and two different audio tracks for smooth transitions. For more information, see:

http://www.asymetrix.com/products/dvp

CONCLUSION

Today's multimedia software will enable you to perform many different video tasks with ease, most of which were previously impossible or were very expensive just a few years ago. Best of all, the software is flexible. You can now piece together the perfect video production with a relatively inexpensive computer system or you can watch video clips of your grandkids from thousands of miles away.

9

HDTV

Everybody knows that HDTV is the future of broadcast television, but few people know what HDTV is or, more importantly, when this potentially revolutionary technology will be making its debut. Although everyone seems to be talking about it, only a handful of people are actually making the decisions that will make HDTV a reality for North American viewers. Part of this confusion has resulted from the many political and commercial smokescreens of this technology.

To recap some information from Chapter 2, *HDTV* is *high-definition television*. From the name alone, you get the impression that the new form of TV will have a more-detailed picture. What you don't see from the name is that it will be all digital, it might provide CD-quality audio, and it will require less power to cover the same general area with a sharp, strong signal. Also, you don't see the long, winding path that the format has taken or the political bickering that has held this technology at bay for years.

Even though a timetable and some specifications for the HDTV system have been assigned, the only definite point about the broadcasts that we can be sure about is this: We have no idea what will happen.

History

Technically, high-definition television is any form of TV broadcasting that has a better resolution than NTSC. It could be an analog system based on the NTSC requirements, but with twice the resolution of NTSC. Or it could be a digital system with MPEG-2 compression and audio based on the CD recording techniques, so long as the video is better than the resolution on the NTSC standards. So, HDTV is not a particular format, like NTSC or MPEG-2, but an entire category of technologies.

The original HDTV goes back to the early 1970s in Japan. The Japanese HDTV format doesn't at all resemble the ones currently being considered in the United States. Modern HDTV in the United States is entirely based on digital broadcasting and compression techniques. Japan's system was a higher-resolution analog version, based on the good ole U.S. NTSC system. Still, the Japanese system was revolutionary because it was the first to dramatically improve upon the original broadcasting standards that had been in place for decades. Detractors have stated that the system was primarily established out of nationalism, isolationism, and greed, but whether or not the allegations were true, the final result was a vastly improved picture.

The rest of the world ignored Japan's HDTV accomplishments until the late 1970s, when Europe began to work toward a new HDTV system to replace their aging PAL system. It never panned out. Despite leadership roles from electronics giants, such as Philips, the fight for HDTV ended in Europe, with the major constituents squabbling with each other instead. Considering the technological advances, it shouldn't be too long before new standards are set into motion again. Based on the past history, however, it's anybody's guess how long it will take Europe to adopt an HDTV replacement standard.

The United States waited until the late 1980s to decide that HDTV could be a future possibility. In 1987, the Federal Communications Commission issued a Notice of Inquiry concerning the possibilities of HDTV. An FCC Notice of Inquiry amounts to an information request. By issuing such a document, the FCC essentially said, "We are interested in changing to HDTV; lobby us with your ideas, complaints, and technological innovations."

If the country with the richest consumer base was planning to switch over to a new standard, you can bet that every major broadcast manufacturer would fight for a piece of the riches. And they did. By 1989, 15 different companies had submitted 20 different possible systems for use as *the* HDTV system. These systems ran the entire gamut from analog to digital, 4:3 to 16:9 aspect ratio, and nearly every imaginable resolution.

In just a few years, billions of dollars in research had been spent with no forseeable results. What was the motivating factor for such spending and such competition? Whenever a new technology is developed, the company with the patents has the right to market it as it pleases. No other company has the right to sell proprietary technology without permission. A great example of this type of deal occurred during the second radio boom in the early part of the 20th century. When Edwin Armstrong invented FM radio, he refused to sell his patents to anyone else. Instead, any company who wanted to build FM radios had to purchase a license from Armstrong, earning him millions of dollars. In a more-recent scenario, Apple Computer operated for years without allowing anyone else to use their technology. As a result, if you wanted an Apple-compatible computer, you had to purchase one that was manufactured by Apple. Recently, Apple has licensed its technology to other companies.

So, if a company could develop the HDTV technology that would be approved by the FCC, they would have complete control of the North American television market. Not only would they have the power to license the technology to manufacture TVs, but also transmitters. At stake was billions of dollars, enough money to keep even a floundering giant company afloat—and flourishing—for decades into the future.

In addition to this commercial battle, there are high stakes for governments around the world. For example, imagine the consequences if the winning system came from a U.S.-owned company that manufactures TVs in the United States. Currently, most TVs are manufactured in China and some other Asian nations. If this scenario occurred, not only would the U.S. government pull in much more money in taxes, but they could also significantly improve their unemployment, GNP, and ratio of exports to imports. Such a ruling would be somewhat similar to suddenly rendering all Japanese and German cars "incompatible" with North American roads! On the flipside, the other countries from around the world could also be dramatically affected by the FCC's decision. If either a company from North America or Europe came up with the winning system, then Japan and their consumer electronics industry would be adversely affected.

Considering the significance of the decision, it seems reasonable to think that the FCC would have looked for the first U.S.-developed HDTV system that was a reasonable improvement and quickly approved it. This way, the FCC could have made a decision that would have dramatically improved the national economy and the quality of TV broadcasting at the same time. It didn't happen. I'm not sure why the FCC didn't take this route because it would have quickly eliminated their own extra work from all of the governmental deliberations and red tape. Even though it would have produced negative international repercussions (especially in Japan, home of the leading consumer electronics manufacturers), the Japanese set the precedent of licensing a worldwide incompatible system that was created by a national company, so the point would have been moot.

Instead, the FCC decided to "let the market decide," a very costly move for the FCC, the companies involved in the technological process, and the public. The past 10 years have been mired with technological releases, system updates, experimental HDTV broadcasts, corporate mergers, and general bickering. After years of this process of consolidating the different systems, we started to get a clearer picture of what the HDTV system would be. The major manufacturers created a group known as the "Grand Alliance" and they decided that it would be a whole bunch of different formats; the TV receivers would be made with several different chipsets so that a a single TV could receive several different formats of HDTV. Just as this ramshackle agreement was beginning to solidify, a consortium of computer manufacturers (including Apple, Microsoft,

Intel, Hewlett-Packard, and others) challenged the Grand Alliance with a totally different system, based on the standards of the Internet and the computer industry. This totally unbalanced what progress had been made over nearly a decade of deliberations.

To put the U.S. HDTV process into perspective, when HDTV was first announced as a possibility, the Internet was not available for the public, a Tandy 6000 computer with a 68000 chip (which offered performance slower than an 80286) cost $5499, the CD (let alone the laserdisc or DVD) was still a novel technology, DBS broadcasting was still years in the future, Beta VCRs were still commonly available new, Michael Jordan had not yet won an NBA championship, and Tiger Woods was in junior high. The process has been ridiculously slow and costly.

THE PRESENT

HDTV is scheduled to be broadcast in early 1998, as this book is being published. The FCC has mandated that all current TV stations must switch over to digital broadcasting by the year 2006. In the mean time, the FCC will begin licensing TV stations to broadcast a second digital channel in the area between channels 60 through 69. So, many stations would be transmitting with one NTSC signal in the standard VHF or UHF TV broadcast band and the same programming simulcast in the new HDTV band. After the transition from NTSC to HDTV, the stations would broadcast in the area that is currently channels 7 through 51 (or possibly 2 through 46). The rest of the TV broadcast spectrum (channels 2 through 6 and 52 through 69) would be assigned to public safety uses and the rest sold in government auctions to the highest bidder.

Despite the glowing reports coming from the magazines and newspapers about the wonderful HDTV technology that we will soon be served with, a number of ambiguities and conflicts point to the contrary.

* *Programming* Plenty of current TV programming is recorded on 35-mm film, but others are recorded according to

NTSC standards. Some programming will not be available in HDTV formats, and everything else will cost extra money to convert from film to HDTV.

* *Equipment* Currently, there is not enough equipment that operates according to high-definition standards to go around to all of the stations that want it. This situation could be alleviated after several years, but it could make HDTV broadcasting impossible for more than a few stations early on.

* *Cable, satellite, and DBS compliance* The cable companies, satellite broadcasters, and DBS networks must switch to a compatible form of HDTV in order for terrestrial HDTV broadcasting to be feasible.

* *Transition* Although a timetable has been established for switching NTSC over to HDTV, very little information is available about how this will actually be performed.

* *Standardization* To date, no standardization of HDTV signals has occurred.

* *Willing compliance* To date, many organizations are not willing to switch to HDTV.

This is a hefty list of potential complications for a broadcasting system that is scheduled to be established right now. In fact, a major development, "willing compliance" occurred several weeks previous to this writing. ABC and CBS led the way with a petition filed to the FCC to overhaul their proposed assignment of HDTV channels. To further complicate matters, NBC and Fox refused to sign the petition. Until the broadcaster networks agree to allow the FCC to assign channels, the switchover to HDTV cannot begin. With the length of deliberations already begun, this setback alone could easily push the start of HDTV broadcasting to 1999 or later.

The Future: The Monster of Progress vs. the Red-Tape Plague

The future for HDTV could range from immeasurable success to financial disaster. Fortunately, it's not an all-or-nothing situation; unlike the DCC audio tape, the amount of success could lie somewhere in between these two extremes.

The HDTV has several very powerful positive components that will clinch its success either according to the FCC's timetable or somewhere in the future. Everyone says that the differences between HDTV and NTSC are astounding. The 50 years of video innovations since NTSC was implemented have truly made the format obsolete. In addition, digital video signals compressed with MPEG-2 techniques will reduce the amount of necessary spectrum width for HDTV to be broadcast. That means that a noninterlaced 1280-x-720 signal with CD-quality audio could be broadcast in about the same amount of space as a traditional analog NTSC station. Or, if the standards were dropped a bit, more stations could fit into the same amount of space that had been allocated for TV broadcasting. As new TV services are developed and the radio/TV spectrum is used for more applications, any "new" space is a tremendous asset. Finally, to answer the question of hardware, we've got the technology to produce HDTV cheaply and efficiently right now. In fact, an HDTV might even be as expensive or even cheaper to manufacture than standard NTSC TV. There's no doubt that some form of HDTV will be implemented. There questions are when and how.

Despite the overwhelming promises of HDTV, its opposition is formidable. The greatest enemy of HDTV is the lack of standards imposed by the FCC. Instead of mandating a specific system or way to broadcast, the FCC has decided to "let the market decide." Now, years later, we are approaching HDTV in a weird commercial free-for-all. Nobody knows exactly what is happening.

A great example of this confusion is the form of HDTV that will be broadcast. There is no single form of HDTV that will be broadcast! Instead, a number of different systems will be used and your TV will have to be able to decode all of them. This means that each HDTV will require extra chips and programming and that you will pay the extra cost, not the FCC or the companies who cannot agree on a single system. And things become more complicated from there. What if part way through the switchover process from NTSC to HDTV, a huge media conglomerate purchases a number of broadcast stations, then transmits those signals in HDTV using a new format that is incompatible with current HDTV standards? It might seem silly to do this, but if that company also manufactured TVs, it could effectively make its competition's TV obsolete, and you would have to purchase either a new TV or a decoder box from that company. The only group hurt by this type of renegade operation would be the viewing public. This type of situation could occur unless the FCC adopts a specific system for TV broadcasting.

Another problem with FCC regulations is their position regarding what can be broadcast and how it could be transmitted. This ambiguity is represented by a number of very significant problems. The biggest problem is that there is no rule that states that the new digital broadcasting must be a form of HDTV! Although reporters have been playing up the "photo-quality" video from the new systems, the FCC has not stated that this must be the case. In fact, in an article in *Wired* magazine entitled "The Neitzen: The Great HDTV Swindle" by Charles Platt, FCC commissioner Reed Hundt was quoted as saying "...let's allow them to use the digital bits for whatever they want..."

The end result is that HDTV might not be HDTV after all. After the decade of hype about the fantastic video and audio quality from "the new system," "HDTV" could just wind up being a digital TV with the same (or slightly better) specifications as NTSC! Rather than being a revolution in the technical quality of terrestrial TV broadcasting, HDTV could wind up being one huge get-rich-quick scheme for the licensed broadcasters and electronics companies.

One question would be how the licensed broadcasters would be able to get rich by transmitting digitally with a lower-than-NTSC resolution. If anything, you would expect their ratings to drop as a result of competition from stations that are actually transmitting HDTV. The problem is that the FCC has stated that the stations will be allowed to use their channel space. If their digital TV signal only uses 15% of their allotted channel space, they are allowed to use the other 85% for whatever they want. These "other services" could wind up being much more lucrative than the actual free TV service that these stations are supposed to be providing for the general public good. As a result, it is very possible that TV stations will be bought specifically for the other channel space. They could transmit 24 hours of reruns of "Gilligan's Island" because it wouldn't matter if anyone even watched the TV station—the important considerations would be to operate the TV station as cheaply as possible and to make money from the other services. If you think that TV programming is bad now, imagine these circumstances!

And just what could all of these "other services" be? Just about any money-making scheme from all-infomercial TV stations, to "video jukeboxes," to pay TV stations, to two-way communications, to computer software to download for a fee, and much more. For the most part, the TV stations could offer the equivalent of junk mail, subjectively good programming for a price, or nonbroadcast programming altogether, all of which could make plenty of money for the broadcasters, but would probably significantly degrade the quality of terrestrial TV.

The "other services" aspect of the new digital TV system is the focus of the scathing article "The Neitzen: The Great HDTV Swindle," which attacks the FCC, Congress, and the broadcasters as kids at a birthday party coming "up with increasingly far-fetched, greedy strategies to grab more than their fair share of the cake." You can check it out for yourself at:

http://www.wired.com/5.02/neitzen

But even more problems lie in the possibilities of splitting the channels apart for "other services." But these aren't just moral difficulties, these are technical. If the FCC divides the spectrum up into different channels, maybe channels 1 through 50, for example, what will happen to the channels if the stations are offering multiple TV programs in their space? For example, if the new channel 23 in your area decides that it won't offer any HDTV transmissions, but will instead offer six different TV stations in its space, then how will your TV receive it? Will there be channel 23A, 23B, 23C, 23D, 23E, and 23F? What if this is the case and the channels are lettered in this manner, but your local channel 23 decides instead that it will broadcast two lower-grade forms of HDTV? These two channels would be of wider bandwidth than any of those in the first example, but yet they wouldn't on the same frequency as any of those. They would be incompatible, unless the TV manufacturers develop software to search for the stations and automatically decode them in the proper format. This would probably be the case, but not necessarily. The complicated technology would add more to the cost of the sets and would likely have some software bugs in the early editions of the HDTV sets. Or, manufacturers might decide that their markets would improve if the less-expensive sets could decode a certain number of channel/format possibilities, the mid-priced sets could receive more, and the best sets could receive everything. Again, the result is a nightmare for the viewing public.

Competition and the sale of digital TVs in the marketplace pose the next problem. For example, what would happen if the satellite/cable broadcasters and DBS networks decided to either stick with their current formats (NTSC for satellite/cable broadcasters and a form of digital encryption for the DBS networks)? If this happened, someone viewing these services would be able to continue purchasing standard TV sets (under the same specifications that are being offered today) and not have to switch to either an HDTV receiver or a digital decoder.

Considering that more than half of the people in the United States and Canada already receive either one of the DBS services or cable TV, these people might decide that they don't need to pay all of that extra money to receive TV. If the general public would show very little interest in purchasing HDTV receivers, that equipment would soon lose its shelf space at the department and appliance stores. If HDTV receivers would fail to sell, at least some of the TV stations would request an extension of their analog broadcasting licenses or stop transmitting HDTV altogether (other stations might make enough money by transmitting nonbroadcast signals that it would make very little financial impact if few people were watching their TV outlet).

Given this scenario, one of five things could happen:

1. The FCC could standardize HDTV operating systems and force stations to switch to one format. After years of "let the market decide," it might be nearly impossible to force such a standard, without years of litigation. In the meantime, the HDTV spectrum would continue to falter.

2. The HDTV spectrum could continue as a niche market, with several million viewers across the nation, rather than having a controlling audience, with well over 100 million potential viewers.

3. The broadcasters could be allowed to return to their analog channels and the FCC could begin planning a new HDTV system for sometime in the future. Of course, the public, broadcasters, and manufacturers would be much more cynical about any new system and it would be less likely to survive.

4. Terrestrial TV could be considered a lost cause and the FCC could sell off all of the frequencies used for TV broadcasting to the highest bidder. Then, the only way to receive TV would be via pay outlets: either cable or DBS.

5. HDTV stations could broadcast in a variety of different formats, allowing the receivers to automatically match the format. In this scenario, chances are that video quality would vary from station to station, with anything from NTSC quality to near-maximum HDTV quality being available in any given major television market.

Just the possibility that in a few years we could be left with only a few pay-TV systems (and no free system) in place is a frightening concept. But, it's not entirely unfounded. The FCC has been very supportive of the pay-TV services in recent years (as is evidenced by the recent creation of DBS broadcasting, work toward a new satellite radio system, and the support for terrestrial pay-TV and other services on the split HDTV services). Also, the recent version of the FCC has been selling the radio/TV spectrum off chunk by chunk to the highest bidder—an unprecedented move before the 1990s. Although it has always been sympathetic to the broadcasting industry, at one time, the FCC worked to regulate what was considered to be the public's airwaves for the best possible public good—by licensing stations and keeping the operations orderly. Now, the FCC seems to be more interested in acting as a broker to sell off what has never belonged to them: the air.

Should you buy an HDTV receiver?

HDTV offers great promise, if implemented properly. Before you purchase an HDTV receiver, be sure that it can receive a great number of TV signals in your area and be sure that the signals are of better-than-NTSC quality. Only consider purchasing one after these points test positively. There are still plenty of uncertainties concerning the whole HDTV system to either condone or condemn it at this point.

Latest developments

Between the time that this chapter was written and the time that the book was about to be published, several notable HDTV developments occurred, but none of these happenings drew us any closer to a completed, operating HDTV system.

The manufacturers have geared up for HDTV and many major consumer electronics companies, including Samsung, Philips, Panasonic, Thomson/RCA, Toshiba, Panasonic, etc., have produced prototype HDTV systems that will be ready to hit the shelves in the upcoming months. All of these models were on display at the 1998 Consumer Electronics Show in Las Vegas, Nevada. The great turnout for HDTV even prompted the *Stereophile Guide to Home Theater* magazine to boldly proclaim that "HDTV is real!" on the cover (Fig. 9-1).

Unfortunately, the vast number of HDTV prototype receivers that are waiting for a chance to hit the market are terribly overpriced for the consumer market. According to the *Stereophile Guide to Home Theater*, the *starting* price for any one of these TVs ranges from $6000 to $11,000. At these prices, it will take years before the costs drop low enough to match the present-day prices for TVs.

Fig. 9-1 The May 1998 proclamation of the reality of HDTV, after the annual Consumer Electronics Show (CES) in Las Vegas, Nevada.

And even though all of these HDTVs are out in prototype form and their pictures look fantastic, still no system standard has been implemented. The TV prototypes will accept any one of the 18 proposed systems. This system is handy because the TVs will work, no matter if the TVs throw you a fastball or a curve. Still, what if the broadcast stations in your area throw a change-up and everything is broadcast in old NTSC quality? You will have paid maybe $8000 for a TV that might barely outperform your old 35-inch model at that resolution. Remember, at the electronics shows, these HDTVs are playing back at maximum resolution, so of course they look fantastic. However, no one really knows at what resolution digital TV will be broadcast to the public. Also, the manufacturers haven't addressed the potential difficulties with stations broadcasting other video signals within the same channel (the channel 23A, 23B, etc., problem).

Most of the preceding problems were addressed earlier in this chapter, but one difficulty was totally unexpected. On February 27, 1998, North America's first HDTV station made history with its

Fig. 9-2 The history-making HDTV station: WFAA-TV in Dallas, Texas.

first trial broadcasts (Fig. 9-2). WFAA-TV in Dallas, Texas, began broadcasting digital signals for experimental purposes only (considering that no consumer HDTVs have even been released yet). The digital TV signals shut down several wireless heart monitors at the Baylor University Medical Center, also in Dallas (Fig. 9-3). I think I can assume that most of the proponents of HDTV hoped that that the medium would be *life enhancing*, not...well, you get the idea !

Conclusion

Perhaps this chapter seems a bit too cynical, in light of the great promises of HDTV. But this is a reaction to the lack of answers that I received during my research and also to the potential problems that bothered me from the information. Also, as mentioned, reporters have only been discussing the huge benefits of HDTV, without discussing any of the potential difficulties. It's always best

Fig. 9-3 HDTV isn't easy: A cartoon making light of the early difficulties of digital television broadcasting. *Monitoring Times*

to temper such optimism with a good dose of cynicism. Finally, be wary of anyone who says that they know exactly what is happening with HDTV. At this writing, no one really has an idea of what will occur. Anyone who says they know is either ignorant or has a stake in the final outcome.

10 Satellite TV

In the early 1980s, many broadcasting groups began sending their signals from outer space to cable-TV stations, which would pay a premium for the ability to offer their broadcasting services to their customers. The concept here was that an entire network could be built around niche programming and people would be willing to pay extra money every month to see it. If enough people watched, they could even sell advertising. As a result, a company with a niche programming concept could establish a full-time network and yet bypass the costly process of buying TV stations around the country.

As you know, cable-TV audiences grew exponentially because of early cable-only stations, such as HBO, The Movie Channel, Cinemax, ESPN, and others. But for every technology, there is a technological hacker. It was no secret that the programming was being supplied by satellite. So, companies began providing special antennas and receivers so that you could watch dozens of channels without having to pay a monthly cable-TV charge. Obviously, neither the satellite stations nor the cable-TV operators appreciated losing revenue to home satellite stations. As a result, there were a few lawsuits against people who watched satellite TV without paying the premium, but it was tough to enforce because some stations were free to view anyway. In the end, satellite TV fell out of favor because so many of the most-watched cable stations began scrambling their signals so that only cable-TV companies with the proper descramblers could view them.

By the early 1990s, many of the treasured satellite receiving systems were being sold for scrap. The technology was viewed by like the Beta movie format—a junky also-ran. But, after a massive publicity campaign, a new digital satellite broadcasting system was established in 1994. Not only was the system advertised as a cool new format, but it was also high-tech: it would feature digital signals with CD-quality video and better-than-NTSC video resolution. The new system has taken off; it is still being heavily advertised, and the viewership is still growing.

But, wait. The new system might be a popular, vibrant format, but the old one is still alive and is also growing!

ANALOG SATELLITE BROADCASTS

It might sound confusing that the sales of new analog satellite receiving systems have almost disappeared, yet there are still plenty of stations on the air to watch. I'm not sure of the figures, but I believe that the number of available programs might have even gone up since the beginning of the 1990s. Between the sheer number of radio and TV broadcasts that you can enjoy and the bottom-barrel prices of surplus satellite-receiving equipment, the medium is worth exploring. First, a look at some of the basic technical aspects of satellite broadcasting is necessary.

TECHNICAL BACKGROUND

Why would anyone really want to broadcast their programming via satellite if the audience is small and declining? Almost none of the analog satellite transmissions are intended for public enjoyment. They are used to link remote broadcast sites to main transmitters or for use as program feeds. For example, if you watch a sporting event, it is being fed to the transmitters via satellite. First, the cameras and the announcers are set up. The announcements and video are directed to a control van or truck, where the video, graphics, and audio are all assembled. The assembled signal from the truck goes directly to the satellite transmitter, known as the *uplink transmitter* because it transmits the signal to the geostationary satellite. The satellite receives this signal from the uplink transmitter and retransmits this back to earth on a different frequency. As you can guess, this transmitter is called the *downlink transmitter* (Fig. 10-1).

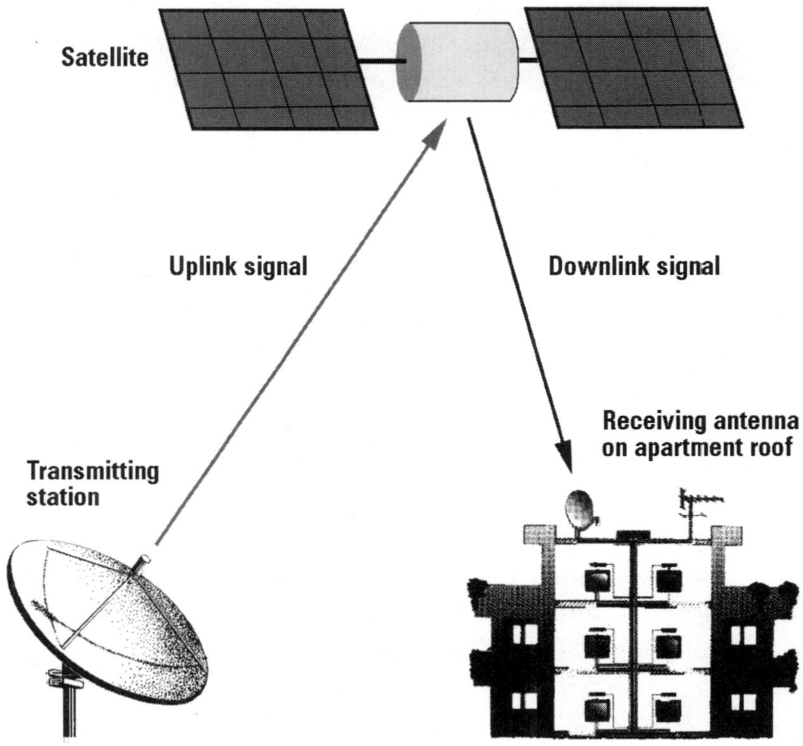

Fig. 10-1 How satellite TV works: The programming is transmitted to the satellite, which receives the signal and transmits it on a different frequency to earth, where it can be received via specialized antennas and receivers.

When the satellite is used as a program feeder, a network, such as CBS, will air all of its taped network programs from a single uplink site. They uplink the programming to the satellite, which then downlinks it. Then, all of the CBS TV affiliate stations around the country can receive the programming with their own satellite receivers and retransmit it.

All of this signal-bouncing stuff might seem like a real waste for the broadcasters, but the opposite is true. The networks experimented with distributing hardcopies of programs at times many years ago, but the extra costs of both the media and the shipping were brutal. And terrestrial broadcasting links are virtually impossible. Chapter 11 on terrestrial television antennas mentions that the low channels provide a much greater signal coverage than the higher channels (given a similar power output and other circumstances). It is simply impossible to cover the United States with a TV signal from one terrestrial transmitter.

The problem with TV coverage is that the frequencies used are line-of-sight. That means that although you don't have to be in visual contact with the transmitting antenna, you must have a clear shot from your TV antenna to theirs. If a mountain stands in the way, forget about it. These signals then zip right through the ionized layers of our atmosphere and off through outer space. Shortwave and AM (mediumwave) signals are reflected back to the earth by these layers, and the broadcasts can be heard around the world.

Satellites effectively provide an "artificial ionosphere" for the extremely high frequencies that television signals use. It's better than the real ionosphere, which varies, depending on the season, time of day, and year.

The next dilemma is getting something up there to retransmit signals back to the earth. Aside from poking around, looking for life on other planets, most space missions involve satellites (whether commercial or military). Even the Space Shuttles launch about one satellite per mission. Many of the satellites from around the world are launched via big rockets. The problem is positioning. That rocket has to accelerate to a speed fast enough to cut past the gravitational field of the earth, yet it has to slow down and deploy the satellite into the proper position. For these reasons, it's easier and safer for companies with enough money to purchase room on the Space Shuttle because they can deploy the satellite gently, at close to the proper altitude.

One point that confused me for a while was how a satellite could provide service to a single geographic area when the earth is spinning and the satellites are in orbit around the earth. The satellite is positioned at a perfect height so that its orbit will exactly keep pace with the rotation of the earth, giving it an essentially stable location. This type of orbiting object is called a *geostationary satellite*. Unfortunately, it is virtually impossible to stick a satellite in one position and keep it there without any help. As a result, the satellites contain little rockets so that they can be repositioned when necessary. One problem with current satellites is that they will fall out of orbit and be destroyed after their finite supply of fuel is depleted.

Another contribution to the death of a satellite is when the rechargeable batteries fail. The satellites are small compartments with a number of transmitter/receivers, batteries, antennas, and solar panels. To power the equipment, sunlight is converted to electricity by the solar panels. This electricity charges the rechargeable batteries. After a decade or so of operation, the batteries' recharging capabilities are depleted and the satellite dies. Length of service for satellites is one of the biggest problems because it is so expensive to send up a replacement unit.

Fortunately, the satellite's transmitters can operate with very little power. In fact, most only put out about 8 to 12 watts (about 1/ 50,000th of the power of a typical UHF TV station). With such a small power output, it is possible to operate a number of transmitters from a relatively small satellite and to work from those batteries for a much longer amount of time (Fig 10-2). It's remarkable that this low power can cover such a large territory with clear signals, but with perfect line-of-sight conditions, very little power is necessary.

Because of the low power, the beaming signal from the satellite, and the line-of-sight signal, each satellite transitter has its own distinctive coverage area. This pattern is often called a *footprint* (Fig. 10-3), and most vary in size from about half a continent to about two continents (for solid signals). Most readers of this book will be in either North America or Europe, where most of the important English-language services are provided. Also, many of the prime stations duplicate their programs for key regions. In other

Fig. 10-2 Scientists prepare a broadcast satellite from the Deutsche Welle (Voice of Germany) for its impending trip to space. Deutsche Welle

words, you shouldn't have to worry about much of the programming being out of your viewing region.

THE BANDS

The two major broadcasting bands for analog satellite TV are the C band and the Ku band. Both have a few advantages and disadvantages.

The C band runs from 3650 to 4190 MHz, and it was the prime band to watch in the 1980s. The band contains scrambled signals from The Sports Channel, HBO, Cinemax, Home Team Sports, The Family Channel, The Disney Channel, Discovery, A&E, Nickelodeon, and many others. Some of the unscrambled options include The Sci-Fi Channel, Storyvision, TBN, BET, The Learning Channel, C-Span, E!, The Nostalgia Channel, QVC, The Travel Channel, NBC, FOX, The University Network, Nebraska Educational TV, CNN Headline News, American Family Entertainment, and much more. In addition to the unscrambled programming that you can find, plenty of radio stations are retransmitted via satellite. This book is dedicated to home video, so satellite audio is skipped here. But if you'd like to find out more about the subject, read *Satellite Times* or call Grove Enterprises about their line of satellite radio receivers at 1-800-438-8155 or

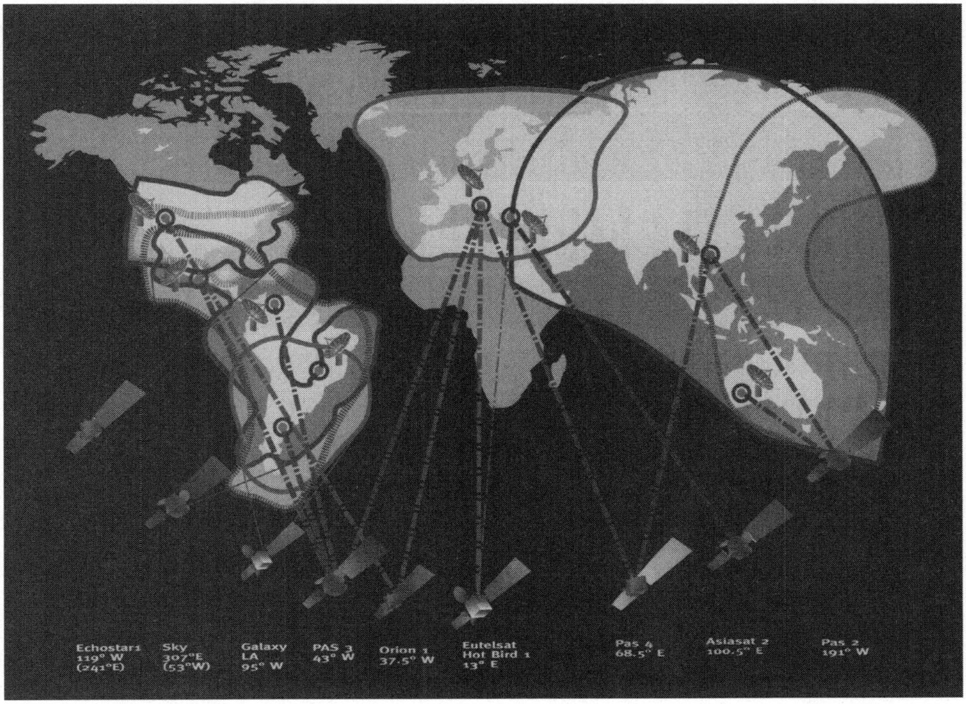

Fig. 10-3 The signal-coverage footprints of various broadcast satellites.

704-837-9200.

C-band satellite signals are easier to tune in because of the larger antennas and the lower frequency. In order to receive a clear signal from a satellite TV station, you need to point your receiving antenna directly at the satellite in the sky. Your antenna focuses the signal to a point, where it is later amplified (more about that in the next section). The overall effect is that although the antenna must be pointed at the satellite, you still have a few degrees (but not too many) to play with in any direction. The big problem with receiving C-band satellite broadcasts is the giant antenna. Who wants to have a 10-foot dish in their yard? Not many people, and that has been one of the major sucessful selling points of the digital broadcast satellites (DBS), which use tiny 18-inch dish antennas.

The Ku band is a bit newer than the C band, and it runs from 11724 to 12182 MHz. Although the activity is picking up on the Ku band, it is still not up to par with the C band. Some of the scrambled stations on the Ku band include WTBS, Showtime, ESPN, CNN, The Family Channel, and others. Some of the unscrambled programming includes that from BBC (UK), NHK (Japan), The Kentucky Network, NBC, Oregon Educational Network, National Weather Network, The Classroom Channel, Youth-TV, Saskatchewan Commercial Network, CBC (Canada), and others. There are also a number of audio-only channels on Ku-band satellites, but not nearly as many as on the C band.

As mentioned, many of the most popular stations broadcast via satellite are scrambled. Although there are still a number of stations (albeit of much lesser interest to most viewers) on the air that would make owning a satellite-receiving system worthwhile if you could find one surplus, part of the interest of many present-day users is in the offbeat. If you turn the dish toward the edge of the horizon, will you be able to pull in a snowy picture from IRIB TV-1 from Iran on 10990 MHz? Or maybe The Cartoon Network broadcasting in French on 11023 MHz? (Fig. 10-4)

The big plus for owning a Ku-band system is that you can receive the video with a much smaller dish antenna. This way, you won't have an ugly monstrosity sitting in your back yard. And if you plan to move, the Ku-band dish is small enough that you can take it with you—without having to disassemble a project that seems nearly as overwhelming as taking apart the Eiffel Tower. Also, if you already have a C-band dish, you can probably use it to receive Ku-band signals (although you will still need a different, frequency-specific receiver and LNB). On the downside, there are many fewer programs to receive on the Ku band, and you need to have the antenna pointed *exactly* at the satellite. Anything less will deliver static.

Better yet, satellites are often used to provide live video from different news and sporting events. Typically, everything goes out live and the commercials and other material are cut in at the main control center. So, with any luck, you will see announcers warming up at the scene of an event, find out what an announcer *really* thinks of the basketball game that he's covering, or see some

Fig. 10-4 Although not bizarre, the Deutsche Welle is one of the many international broadcasters that you can receive via satellite.

walloping bloopers. Some of these blooper home recordings from satellites have been passed around in the video underground for years. With a cheap system, you might be able to catch some of this action firsthand.

An even more esoteric branch of satellite monitoring extends firmly into the hobbyist realm. Some people make a hobby of tracking satellites that have been "left for dead." These are old satellites that either ran out of fuel or are no longer maintained. As a result, they are no longer in a geostationary orbit, and they zing around the earth, basically unnoticed. In some cases, the batteries have not died and the transmitters are still pumping out blank signals. So, with a lot of monitoring, a hobbyist might notice a signal that slowly drifts up or down as a result of Doppler shift. Bingo! In a few minutes, the signal will disappear, only to return in another hour or two. By logging the time between fade in and fade out, they can determine the height of the satellite above the earth. Armed with that information, some satellite charts, and a lot of determination, the satellite hobbyist can sometimes figure out what old satellite is being heard. It might be a peculiar hobby, but it is interesting and potentially helpful to the professional satellite industry.

RECEIVING EQUIPMENT

Because of the differences in frequency between satellite broadcasts and standard terrestrial TV broadcasts, you will need a different antenna and receiver to connect to your TV. Many companies used to sell complete satellite TV receiving packages, but these days, much of the available equipment is used or surplus, so you might wind up piecing a receiving station together. As a result, this section covers the necessary components piece by piece.

ANTENNA

Unlike antennas for most different radio or TV broadcasting services, the satellite-receiving antenna consists of different important components. For example, if you wanted a TV antenna, you would buy a TV antenna. You wouldn't buy the TV antenna framework, buy some specialty VHF elements, and finally buy a UHF antenna to complete the array, but this is what buying an analog satellite-receiving system is like.

The components of a standard satellite antenna include:

* *The dish* Satellite receivers essentially require a dish-type antenna to focus the signal. "The dish" (also known as a *parabolic reflector* because of its shape and job) received it's name because of

its likeness to that particular container. Most dishes for analog reception are perfectly round and the signal is focused to a point directly in the center. A dish antenna can be of any size, but the size determines its effectiveness. A larger antenna will be able to bring in a stronger signal, and it will also be less susceptible to reflection errors (such as if the dish fills with snow) or being turned slightly off a straight line to the satellite. Most C-band dishes are typically anywhere from 8 to 10 feet in diameter, and most Ku-band dishes are approximately 2 to 3 feet in diameter.

Another consideration of the dish is its construction. Most C-band dishes are constructed from screwed-together sections of either flat,

Fig. 10-5 A side view of a C-band antenna, showing its screen and frame construction.

painted metal or metal screening (Fig. 10-5). Most of the small antennas are constructed from a single piece of metal. The advantage of the one-piece system is that it is very resistant to warpage, bending, or other damage, which could render the focusing power of the reflector ineffective. Basically, the more sections that an antenna consists of, the more likely it will become misshapen. If you are thinking about buying a used dish, take a look at the signals produced using that antenna. Also, especially if it is no longer used and you can't see those satellite programs rolling in, carefully inspect it for nonlinearities (primarily dents and warpage). If it is messed up, don't buy it unless you think that you can fix it *and* the price is great.

* *The feedhorn* In most areas of electronics, this device is called a *waveguide*. The feedhorn's duty is to guide the radio waves in a certain direction. It is always used at very high frequencies and acts kind of like a funnel for radio-TV waves.

* *The polarizer* The polarizer performs a confusing little job. It is a small device that fits between the feedhorn and the LNB, or it can be built in a single unit between those two components. Radio-TV signals (waves) can be either horizontally, vertically, or circularly polarized. At radio frequencies, it doesn't really matter which way they are polarized (often even the ionized atmosphere changes a signal from being vertically polarized to horizontally, or vice versa); you can still receive the signal just fine. But, radio-TV theories sometimes act a bit weird in the high frequencies that are used for transmitting satellite information. As a result, you must have the signal aligned properly with the LNB. The polarizer matches itself with the signal and outputs a signal that matches that which the LNB is aligned for. Some polarizers aren't capable of tuning signals at a certain polarity and others do it automatically. The polarizer is one of the important, and confusing, aspects of receiving satellite TV signals (Fig. 10-6).

* *The LNB (low-noise block)* This unit goes by a number of different names, but it is most of the big block that you see in front of the satellite dish. The LNB consists of two basic electronics components: an amplifier and a frequency converter. The amplifier raises the signal level to an amount high enough that it can make it

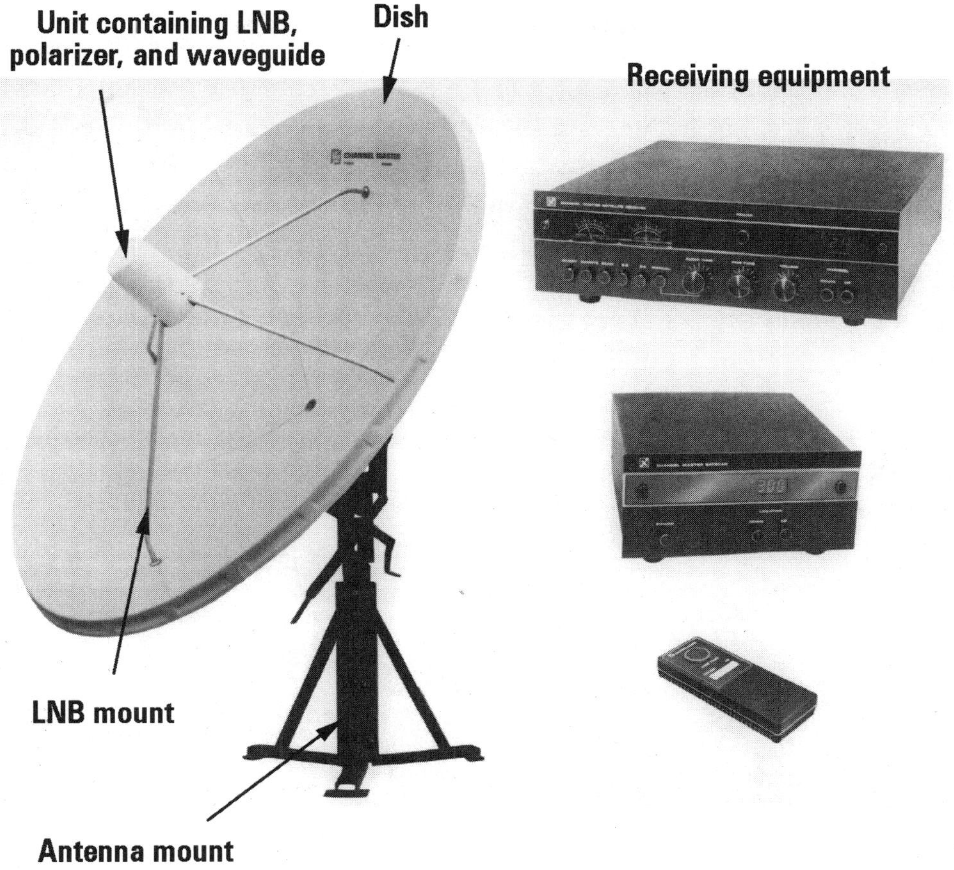

Fig. 10-6 A complete Channel Master satellite TV system from the 1980s.

back to the receiver and provide good video and audio. The frequency converter lowers (downconverts) the frequency of the signal so that much of it isn't lost on its way back to the receiver. For more information on this process, see the entry for *coaxial cable*.

* *The LNB mount* The LNB mount is just a single rod or beam or a group or rods that is used to mount the LNB, polarizer, and feedhorn in front of the receiving dish.

* *Coaxial cable* Most people are used to coaxial cable not making much of a difference. You just pick up something that's

really cheap and use it. You can't do this at satellite frequencies. The keys to good coax are copper braiding with as much coverage as possible and a noncontaminatable dielectric (insulator). Avoid coax with foil shielding inside. Think of the cable as being a water hose. At such high frequencies, the signal will leak out like water out of a sieve. Also, the coax must have the same impedance as your satellite receiver's input. The safest bet is a military grade of RG-6U cable. If you can't find mil-spec coax at your local electronics store, try calling Cable X-Perts at (800) 828-3340 or (708) 506-1886 if you live outside of the United States.

* *Receiver* The heart of the satellite-receiving system is the receiver. The satellite receiver functions just like a TV receiver (except without the screen) or a radio. Its purpose is to do something with those signals that are pulled in by the antenna and

Fig. 10-7 Two different satellite receivers from Drake, which has a long-standing good reputation for high-quality equipment. Drake

delivered by the coaxial cable(Fig. 10-7). Receivers do vary in quality. A good receiver is very *sensitive*, meaning that it can makes sense of very weak signals. In addition, the receiver must also be extremely *selective*, meaning that it must be capable of keeping different signals apart (not receiving several different frequencies at the same time). Selectivity and sensitivity work against each other (an engineer can easily make a receiver more selective by reducing the sensitivity, and vice versa), so a great receiver must have both

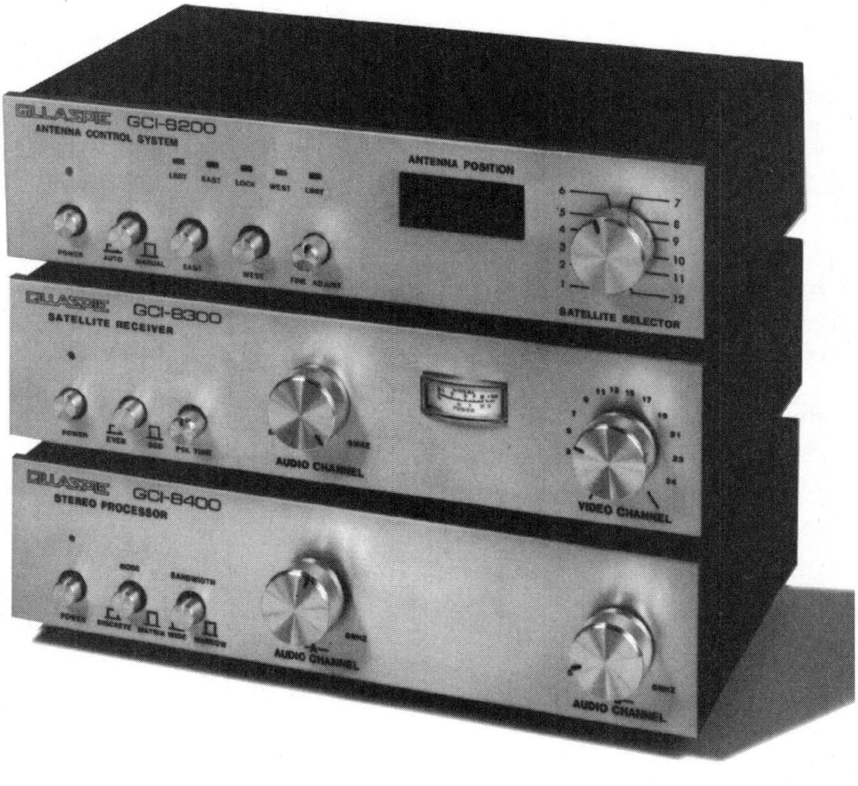

Fig. 10-8 This 1980s satellite receiver is divided a bit differently, with separate components for the antenna controller, the receiver, and the stereo processor.

qualities (Fig. 10-8).

As mentioned, these days, very few large companies are selling new satellite equipment. Grove Enterprises is offering satellite radio receivers, but little else in the the line of satellite equipment. A number of companies sell satellite descramblers, but these are not intended for the private descrambling of video signals. Two of the few mail-order companies still left selling a full line of equipment are Satman, which can be contacted at (800) 472-8626, and Skyvision, which can be contacted at (800) 334-6455. The R.L.

Drake Company has a great reputation as a long-term manufacturer of satellite receivers. If you are interested in finding a local satellite dealer, call them at (513) 746-4556 to see if they have one near you. Or check their Web page at:

http://www.rldrake.com

To find some used equipment on-line, try the For Sale Message Board on DBS Dish.com at:

http://dbsdish.com/message/forsale.html

This board mostly covers DBS equipment, but some C- and Ku-band equipment is available here.

PROGRAMMING

The only major viewing problem stems from the fact that most of the companies that operate these media channels don't want private individuals tuning their signals in. So, to discourage intruders, they scramble their signals. You could bypass their attempts to hinder your viewing pleasure by using a descrambler, but this is currently illegal. So, the unscrambled video offerings aren't prime territory, but there's still some entertaining stuff out there.

INFORMATION

Considering that analog satellite TV viewing is a real niche market these days, it's no wonder that information about the topic isn't flooding mainstream newspapers and magazines. But, with a little effort, you can find plenty of good information. In the early 1980s, there were a number of books on the suject, but these are now out of print. One superb annual reference is the *WRTH Satellite Broadcasting Guide*, published by the editors of the *World Radio TV Handbook*. For more timely information—especially if you make a hobby of using your satellite receiver—try *Satellite Times* magazine, which is published by Grove Enterprises (see their phone number earlier in this chapter) or view their Web page at:

http://www.grove.net
Speaking of the Internet, a few good resources are there, too. Aside from the *Satellite Times* page, try Robert Smathers' page at:

http://www.nmia.com/~roberts/robertext.html

It isn't much much to look at, but it is loaded with plenty of helpful information. One excellent provider of interesting downlink frequencies and satellite launching activities is Launchspace.com at:

http://www.newspace.com/ref/html

If newsgroups are your thing, try taking a look for information or for sale classifieds in:

rec.video.satellite.tvro

DBS TV

Analog satellite TV is interesting and there's some neat stuff to watch, but it's more of an experimenter's field than anything. Not only do you have to piece your equipment together, but you have to deal with scrambling and switching between dozens of different satellites (i.e., changing the direction of your antenna). This atmosphere provides a great hunting ground for the hobbyist, but it's hardly conducive for viewing by someone who simply wants to plop down in front of the TV and catch a show.

DBS (direct broadcast satellite) broadcasting was created specifically to fill this void. Although DBS uses the same general technical format as analog satellite TV, the approach is totally opposite. DBS is intended for noncommercial viewers—typical people who just want to watch TV. So, rather than analog satellite TV, where the programs are all intended for TV stations, radio stations, or cable companies and the general public is expected to stay out, DBS reception is made as easy as possible for you.

Essentially, DBS was created as an alternative to cable TV. The idea was that if a small, inexpensive system could be developed to receive prime channels that are typically only available from cable TV, plenty of people in remote locations would jump for it. Better yet, if the the signal could be delivered at better-than-NTSC standards and the cost was competitive with cable, plenty of people who presently receive cable would switch over for the better-quality picture. So far, it's paid off and the DBS systems are flourishing across North America. But first, here's some technical background information on the format.

TECHNICAL INFO

Direct broadcast satellites operate in a manner that is similar to analog satellite TV, but there are a few key differences. Like the analog broadcasting that occurs in the C band and the Ku band, the first important factor is the satellite. The programming is transmitted from the uplink center to the satellite. The satellite receives the signal and retransmits it back to earth on a different frequency in the Ku band.

One of the major selling points of the DBS TV services is that only 18- to 24-inch (in diameter) dish antennas are required to receive very clear signals. The main reason is that the on-board transmitters use much higher power levels (about 100 watts output) than traditional analog satellite transmitters (about 8 to 10 watts output). This is obviously much more attractive to someone just getting into satellite TV viewing.

Because these broadcast services are provided without scrambling via satellite for individual viewers, there has to be some way for the the companies involved to make money. The satellite providers make their money digitally. Because the systems are digital and the receivers are likewise digital, the receiver must be digitally unlocked to provide service. As soon as you subscribe, data will unlock your receiver and allow your system to receive programming. But as soon as you stop paying for your service, your receiver will be blocked and you won't be able to receive anything. Likewise, if you subscribed to a system and your equipment was stolen, you could place a call to your DBS provider and they could digitally block the equipment, rendering it useless. The digital capabilities of the system are also neat because if the system or compression is changed, the company can send new software via their satellite to reprogram your receiver—like the opposite of planned obsolescence.

As mentioned, the signals from DBS satellites are digital, not analog. PrimeStar, the first DBS provider, originally offered analog signals when it made its debut in 1994, but it soon switched to an all-digital format as competition increased. The digital formatting for DBS broadcasters is based on the MPEG-2 compression standard. MPEG-2 probably sounds familiar because it is the same standard that is used for DVD and HDTV. As mentioned in those chapters, MPEG-2 is so sucessful because much of the coding is based on a change in data, rather than what each bit of data is.

With the MPEG-2 format, a given piece of video can be reduced to about 16% of the original data requirements (or less, depending on the amount of change in the picture information). The digital compression not only makes a better picture possible, but it allows more channels to be shown and lowers the subscription cost of the system. With the MPEG-2 compression, as many as six channels can be shown per transponder. With analog transmission, only one station can be shown per transponder. A typical broadcasting satellite can now deliver 100 channels (or more) of programming, as opposed to 16. Thus, you receive more programming and the satellite company doesn't have to pay the extra money to launch several satellites to provide a decent service.

The important thing to remember is that *DBS* means any digital satellite service that broadcasts directly to its viewers. PrimeStar, DirecTV, USSB, and EchoStar are all companies that offer different types of DBS formats and systems. DBS systems are not all compatible with each other. As a result, if you have the PrimeStar system, you will not be able to receive programs from the other companies' satellites. Likewise, if you bought used AlphaStar receiving equipment, you would not be able to receive programming from any other system. Because AlphaStar, an upstart DBS service that lasted less than a year across part of 1996/1997, went bankrupt, never buy any used AlphaStar equipment! It's now useless.

COSTS AND PROGRAMMING

As mentioned in the last section, each DBS company offers different equipment to decode broadcasts from different satellites. Although they all use the MPEG-2 coding systems, the transmissions are not compatible, except for DirecTV and USSB, which have a partnership. As you might expect, the different companies are also scrambling to provide more attractive prices, programming, or packages. What you might decide to purchase depends on what stations you want to see, the cost of the equipment, and the monthly charge.

For example, just a few years ago, DBS systems were selling for $900 apiece, plus the monthly fee. EchoStar, the newcomer to the DBS block, began severely undercutting the equipment prices of the other companies, forcing them to also cut costs. As a result, the companies are competing on all sorts of different levels. One of the competing fields is price per package. EchoStar, the original cost cutter, is offering a few different packages at this writing, which reflect the full spectrum of channels per dollar:

* For $3.95 per month, you can receive three network broadcast stations and two network feeds (ABC, NBC, CBS, FOX, and PBS) from either the East or West (provided that you meet the FCC criteria regarding receiving network broadcasters via DBS systems).
* For $5.95 per month, you can receive both packages of East and West broadcast stations (10 total).
* For $19.95 per month, you can receive what EchoStar calls its "Top-40" program; 40 of the most-requested satellite broadcasters in one package. This includes Nickelodeon, CNN, The Nashville Network, MTV, The Discovery Channel, ESPN, ESPN2, two Disney channels, The Weather Channel, Comedy Central, and much more.
* For $15 per month, you can pick 10 different stations from the "Top-40" program (except for Nickelodeon, MTV, VH1, and a couple of others).

* For $24.95 per month, you can receive what EchoStar calls its "Top-50" program, the 40 stations from the Top-40 program, plus 10 others.

* A number of pay-per-view (PPV) and other channels are also available.

Really, the only way to know what programming you want to get is to contact each company and receive information on their latest packages and prices:

http://www.primestar-tci.com	Primestar	1-800-PRIMESTAR
http://www.directv.com	DirecTV	(800)-DIRECTV
http://www.echostar.com	EchoStar	(800) 333-DISH
http://www.dishnetwork.com	EchoStar	
http://www.ussbtv.com	USSB	(800) 238-8378

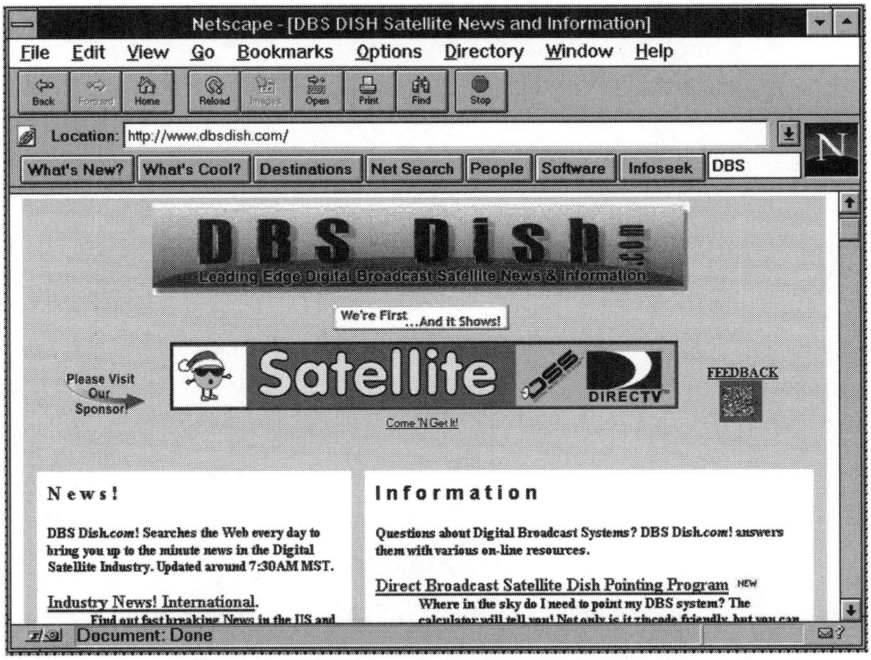

Fig. 10-9 DBS Dish.com on the Internet.

Another sure-hit site is DBS Dish.com (Fig. 10-9), which features news, links to on-line DBS TV schedules, a chat room, a DBS FAQ, and much more at:

http://www.dbsdish.com/

If you want to see plenty of talk about the different services, likes/dislikes, complaints about billing, and other personal experiences with DBS, see:

rec.video.satellite.dbs

EQUIPMENT

Receiving a DBS station is very similar to receiving a station via Ku-band satellites (like a smaller version of receiving C-band broadcasts). One difference is that everything has been made smaller and easier. All you need is a receiver, the antenna assembly, and the coaxial cable that connects everything together. Just consider it as a "plug and play" unit (Fig. 10-11).

Fig. 10-11 A complete JVC DBS receiving system for the Dish Network. JVC

As mentioned, just a few years ago, prices for DBS satellite equipment were around $900. Now, many of the complete units go for about $250 to $400, although the current rebates are often dropping prices down to $150 or less. I've seen EchoStar systems selling for as low as $70 + $20 for the installation set. PrimeStar takes an alternate approach—they don't sell their equipment, but add the price of their equipment rental into the monthly charge.

Aside from buying the equipment directly from the satellite TV provider, you can also try purchasing it used. Probably the best place on the Internet to find a used DBS system is from the For Sale Message Board on DBS Dish.com at:

http://www.dbsdish.com/message/forsale.html

Also, be sure to check the rec.video.satellite.tvro newsgroup.

Installation and Usage

Installing a DBS system is really easy. The companies learned from the bad experiences of many C-band satellite users from the 1980s and early 1990s. You only need to connect the receiver to your TV, then comes the toughest part. You need to find the location of the satellite in your sky and find a clear location where you can mount it. Remember that Ku-band satellite signals are line-of-sight. The line from the antenna to the satellite must be perfectly clear. Even if you have a clear path, but tree branches sometimes will blow in the way, the signal will either fade or "snow out" altogether. Even periods of extreme weather can hamper the signal. DBS is unforgiving.

Once you find a good place to mount the antenna, with a clear shot at the satellite, then you need to find the satellite. This varies from system to system, but DirecTV has a test mode. In this mode, the screen will essentially blank when no signal is being received. When a signal comes in, the screen will show a digital signal-strength meter. When I was installing a DirecTV system for my sister, we had a difficult time finding a good location to mount the dish. It seemed that there were always trees in the way and her landlords wouldn't permit a permanent mounting location. In addition to making the mounting very difficult, it also made finding the satellite tough. After about 30 minutes of pointing the antenna aimlessly at the sky and wondering if this was just some cruel hoax, we happened upon a signal. Then a long session of positioning, checking the signal-strength meter, and repositioning ensued. Finally, we discovered that a 76% signal strength was about as good as we could get and the antenna could only be tilted about one inch in any direction before the signal faded out!

Because of the digital encryption, a 76% signal was still superb—much better than NTSC or VHS signals. With a call to the DirecTV toll-free number, a telephone hookup, and a few minutes of waiting, the system was up and ready to roll.

THE FUTURE

The future of DBS broadcasting is wide open. Right now, it appears that DSS and possibly HDTV (if it ever gets running properly) have the upper hand on cable TV, the first time that cable would have the disadvantage in more than a decade. Right now, the big advantages are that you can receive better-quality audio/video from the DBS broadcasters and that cable TV requires constant maintenance to keep all of the cable runs operating properly. In short, cable TV is more expensive per viewer, which is a big disadvantage.

But the biggest advantage for DBS systems might be coming up within the next year or two. Within this time, DBS systems will integrating with computer data services. The end result will be computer data services, super fast Internet connections, and possibly some interactive TV stations that are only available via DBS providers. Already, EchoStar is providing an agricultural data service for farmers for about $30 per month. Also, DirecTV is providing a satellite Internet connection called DirecPC. With the DirecPC service, you can access the Internet at high speed...but there's a price for the high-speed technology. Current rates are $19.95 or $39.95 per month, depending on the amount of data downloaded. Additional data can be downloaded at 60 cents per meg! At that price, you could quickly build up a tremendous bill!

For more information, see:

http://www.datasat.com/products/direcpc.html

With this service established, it seems likely that the other computer-related services will soon follow.

CONCLUSION

Satellite TV is a bizarre territory, filled with nearly every type of video and audio service imaginable. It includes everything from the mysterious to the mundane, from foreign to local, and from free to downright expensive.

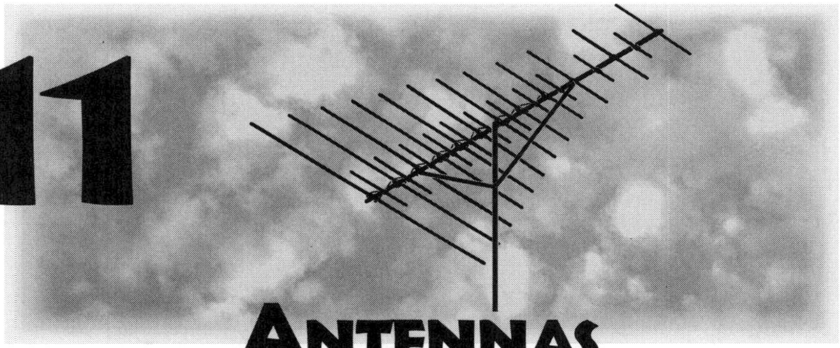

11 Antennas

In the 1950s, the real sign of consumer technological progress was a TV antenna on your roof. The TV antenna (along with the atom, the jet, and space travel) was one of those symbols that separated the past from the future. Now, 50 years later, the TV antenna is obsolete for more than half of those in North America. Still, terrestrial broadcast TV is viewed by millions in North America, let alone the rest of the world, and despite the attractions of cable and DBS TV, it is still the choice of many.

Depending on where you live and what you want to watch, it might be useful for you to work with antennas.

Safety

The first word on masts and towers is *safety*. That's it. If you die or get sued out of your house because of a faulty mast, your antenna won't do you much good. Make sure that your antenna mast is soundly designed, by following good engineering practices, and make sure you follow local ordinances regarding antenna installation.

The primary safety concern for your antenna is that it must not be within falling distance of power lines. For example, don't even think of putting a 35-foot tower within 35 feet of power lines. In fact, it should be at least a few feet further away, in case your estimated distance was incorrect or if something bizarre happened (such as if a hurricane blew your mast over).

Every year, stories abound about amateur radio operators who are electrocuted accidently when they touch power lines while raising a mast or tower. It doesn't matter if you are using a metal mast or something supposedly nonconductive, such as PVC or wood, power lines are still extremely dangerous and should most certainly be avoided.

Before building an antenna or even working on plans for a mast, call the local authorities. In most cases, you can call either the city or boro manager or an inspector who is employed by the township that you live in. That person will be able to tell you what types of antenna towers are permissible where you live. All towns vary with their ordinances, so get complete, verified information. Some historic towns don't even allow outdoor antennas of any sort, so it is up to you to find out.

Last, this chapter is merely a guide to installing your own antennas. Conditions vary around the country and not every antenna mast will be as safe in every location. Thus, it is your responsibility to make sure that your antennas and antenna masts are strategically placed and structurally sound.

A FEW ANTENNA BASICS

An antenna is any object that is used to pick up radio or TV signals. Of course, most antennas are metallic conductors of a certain length, but antennas can be made from other materials. For example, human skin isn't a very good conductor, but if you hold on to a retracted whip antenna (like those used in portable TVs), the signal will improve because you have become part of the antenna.

Part of the reason why radio-TV theory can get very complicated is because it seems so intangible. However, here are a few of the basics. Every radio-TV signal is a sine wave, and every wave that is on a different frequency has a different length. The frequency is merely the number of waves that occur over the course of one second. As a result, a higher frequency yields a smaller wavelength. Knowing that light (and subsequently all radio-TV waves) travels at

300,000,000 meters per second, you can determine the wavelength of a radio-TV wave at any given frequency.

$$Wavelength = \frac{300{,}000}{F}$$

Where:
F is the frequency in kHz
Wavelength is the wavelength in meters

From this equation, you can determine that a full wavelength at 540 kHz (the bottom of the AM broadcast band, also known as the *mediumwave band* in Europe) is 555.56 meters long. At 1700 kHz (the top of the AM broadcast band), a full wavelength is reduced to 176.47 meters. The TV (VHF and UHF) broadcast bands are much higher in frequency than the AM (MW) band, so the lengths of the waves are much smaller. The only difference in determining the wavelength from the previous equation is that you must use the constant of 300 instead of 300,000 so that you can use the frequency in MHz. Thus, at 88 MHz (channel 6 in the VHF TV broadcast band), a full wavelength is only about three meters long. For these equations, the lengths are typically measured using the metric system.

The catch to all of these equations and measurements is that antennas perform at their peak if they are resonant to the frequency range that you are listening to. The resonant antenna is an antenna that is either the same length as or is a multiple or division of a given frequency. The most common antennas that are less than one wavelength long are: 1/2, 1/4, and 1/8 wave. An 1/8-wave vertical antenna for 540 kHz would be 69.45 meters high (approximately 225 feet)! Obviously, a 225-foot high antenna in your backyard is a bit impractical for the average homeowner. So, the radio designers go with smaller antennas, typically either whip or dipole antennas for FM band reception, and loopstick antennas for the AM (MW) band. TV antennas use a complex variation of the dipole antenna, which is covered in the next section.

COMMON FM AND TV ANTENNAS

This section lumps antennas for FM and TV use together because the FM band is located in the middle of the VHF TV band. In fact, VHF TV antennas are more similar in length and design to FM antennas than to UHF TV antennas.

For the most part, TV antennas are considered to be stock components—you just pick one up from a department store or (if you're feeling really high-end) you might stop at Radio Shack to look at their antennas. But, there are relatively few good, high-end manufacturers of FM antennas. One, Ron Smith Aerials (98 Ash Rd., Luton, Beds LU4 8AQ UK), specializes in high-end FM and TV arrays—from 4 to 92 elements! Some of these antennas are truly complex variations on the beam and Yagi themes (see the following sections).

One of the most common types of antennas in use today for FM listening is the dipole antenna (Fig. 11-1). Unlike the simple

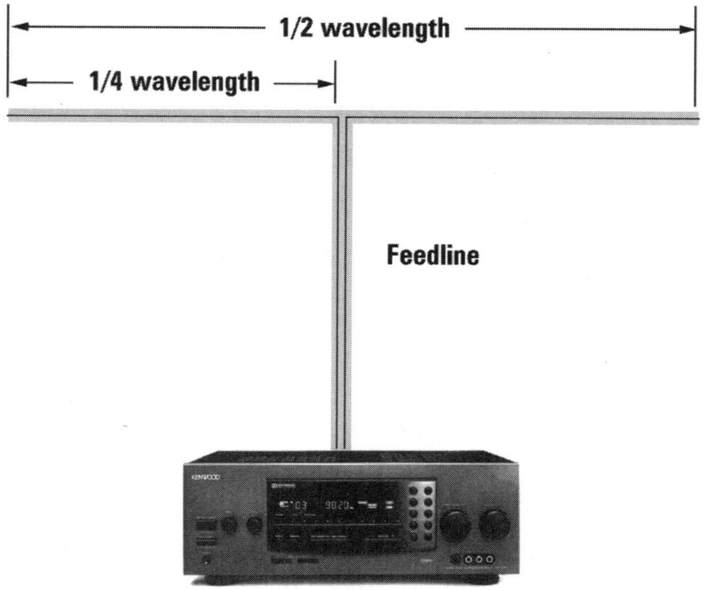

Fig. 11-1 A look at one of the most common antenna types, the dipole.

longwire antenna (which consists of one long, straight piece of wire), the length of the dipole is frequency dependent. That means it must be cut to the particular frequency range (or a clean division of that range) that you are trying to receive or it will be virtually ineffective.

The most commonly used types of dipole antennas have a transmission line that carries the signal from the antenna to the receiver. Either coaxial cable or twin-lead wire is used for the transmission line of a dipole antenna (depending on the antenna impedance of the receiver).

The transmission line is run upward, vertically from the receiver, typically at least 1/4 wavelength above the ground. A 1/4-wavelength piece of wire extends horizontally from one lead of the transmission line, and another 1/4-wavelength piece of wire extends horizontally from the other lead of the transmission line (in the case of coaxial cable, one antenna wire runs to the center conductor, and the other goes to the shield).

Dipole antennas perform best when they are at least 1/4 wavelength above the ground (preferably much higher). So, although you might have plenty of space to install a dipole that is 300 feet long for the top of the AM broadcast band, chances are that you will have a problem raising the whole assembly at least 150 feet into the air. As a result, dipole antennas are primarily used for receiving or transmitting signals from the shortwave bands up through the VHF bands.

Dipole antennas are incredibly popular because they are relatively simple, inexpensive, and effective for general receiving applications. Another advantage is that they are nearly omnidirectional (they receive well in all directions). Because of these advantages, most modern stereo receivers are prepackaged with a little twin-lead dipole antenna that you can stick on the wall. If you don't have one of these antennas, it's easy enough to either make one or pick one up for a few dollars from Radio Shack or from a mail-order company. The difference in reception quality on the FM band between a simple dipole antenna and a long piece of wire is amazing!

DIRECTIONAL TV AND FM ANTENNAS

The advantage of omnidirectionality is that you can receive from all directions at once; the problem is that you can't focus your antenna's signal-receiving capabilities in one direction. If you live far away from any major cities, having a dipole antenna will help you to bring in local stations, but little better than that. If you want to receive stations from more than about 30 miles away, you will need to "focus" the antenna. Because both the audio and the video signals must be brought into near-perfect "focus" to watch TV comfortably, simple dipole antennas are rarely used to receive any type of TV signals.

To "focus" a longwire antenna, you make the antenna longer. That's fine for random-wire antennas on the AM and shortwave bands, but this method simply doesn't work for dipoles on any band. Instead of adding to the length, you can add separate antenna parts to "focus" the antenna. The most common of these "focused dipole" antennas

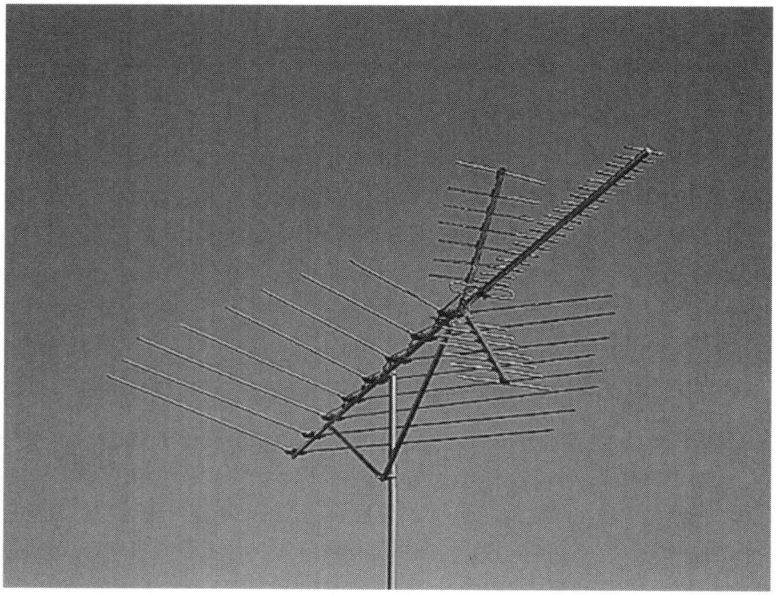

Fig. 11-2 A Yagi television antenna with the antenna elements facing in toward the director elements. This adds to the focusing "power" of the antenna.

is called a *Yagi antenna* (Fig. 11-2). The Yagi features a smaller (by 7%) element to one side of the antenna that "pulls" the signals in (called a *director*), and a larger (by 7%) element to the other side that reflects the signals back (called a *reflector*). The effects that the director and reflector have on the signal reception are interesting because neither is electrically connected to the original dipole antenna (called the *driver* or the *driven element* in a Yagi antenna).

Because of the large size of the antenna assembly, Yagi antennas are typically only used at high frequencies, although very large versions are sometimes used in the upper shortwave frequencies. The size of these antennas, however, is perfect for frequencies in the VHF and low UHF bands. So, Yagis and variations of this antenna are often used for FM and TV reception. Because the antennas are small enough on these frequencies to be rotated, they are always commercially manufactured using light aluminum tubing and installed on a mast.

A three-element antenna will provide significant gain (another way of saying the amount that an antenna has been "focused") over a simple dipole antenna, but the gain can still be dramatically

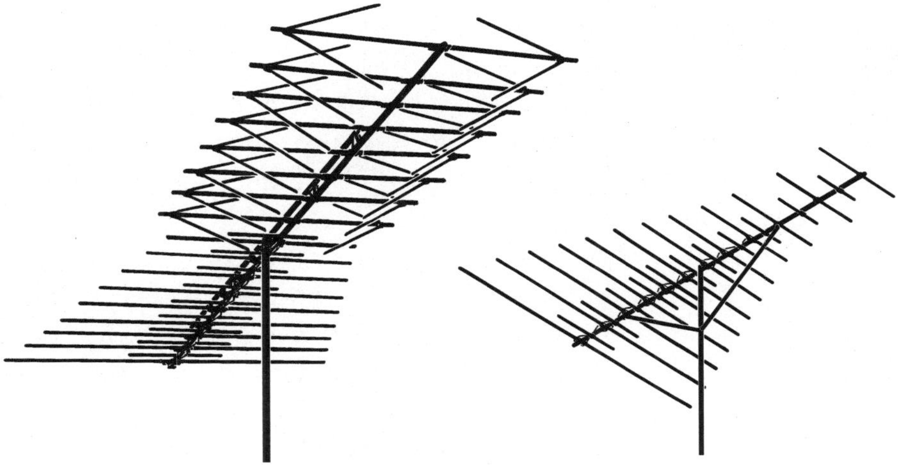

Fig. 11-3 Two different designs for TV receiving antennas. Channel Master

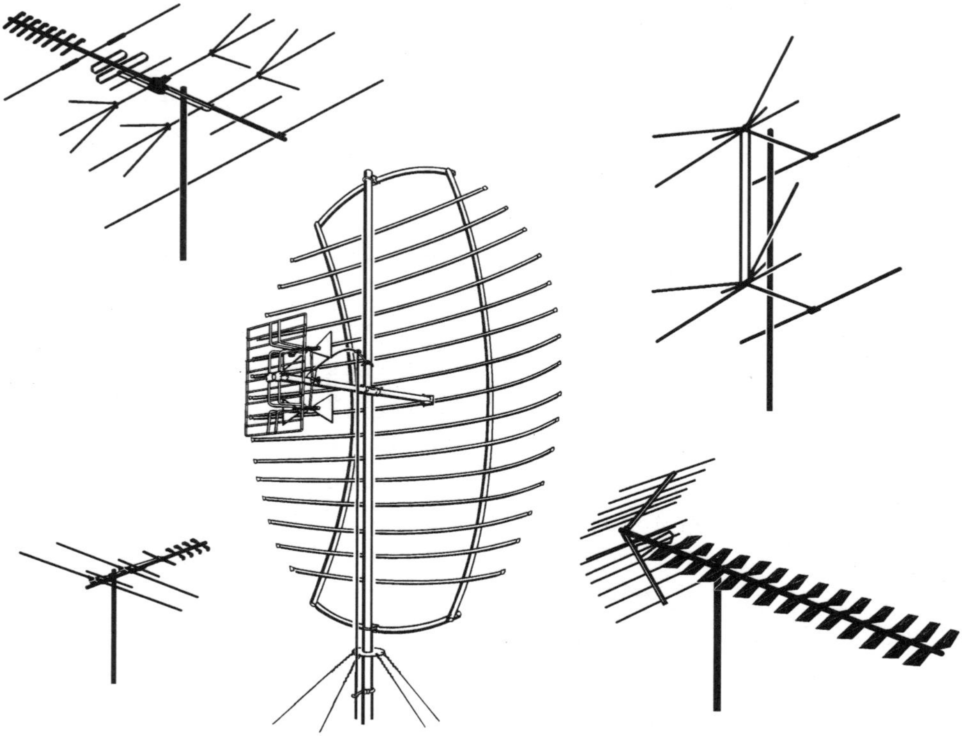

Fig. 11-3 (continued) More TV receiving antenna designs, including Yagi variations, stacked dipoles, and even a parabolic reflector. Channel Master

improved. The easiest way to improve reception is to add more director elements. A good indication of how well an antenna will work depends on its size and the number of elements. A few other "tweaks" that many manufactures use to improve reception are to slant the elements in toward the directors (smaller end) and to stack two sets of antennas.

For example, a high-gain multi-element Yagi antenna will receive a station when you point the antenna directly toward it with the rotor (Fig. 11-3). But, when you turn the antenna away from it, the signal will begin to fade, then disappear altogether. The term for turning an antenna away from a signal so that it can't be received is *nulling out*. So, if you live between two stations on the same channel, you might be able to point the antenna toward one station ("focusing" the signal in) and away from the other (nulling it out). However, because the government regulatory agencies are careful about causing TV interference, it is a rare circumstance where someone is in a location where they can receive two stations on the same channel. This mostly occurs on the low channels (channels 2 through 6) where the signals are capable of traveling much further.

ANTENNA HEIGHT

As mentioned, dipole antennas and their variations perform better at greater heights above the ground and other objects. Part of this is the curvature of the earth, and part is because the VHF and UHF signals are primarily received via a line-of-sight path. In other words, unlike AM radio, you can't receive FM or TV signals if you live in a hollow or if a mountain range is between you and the station that you want to hear (Fig. 11-4).

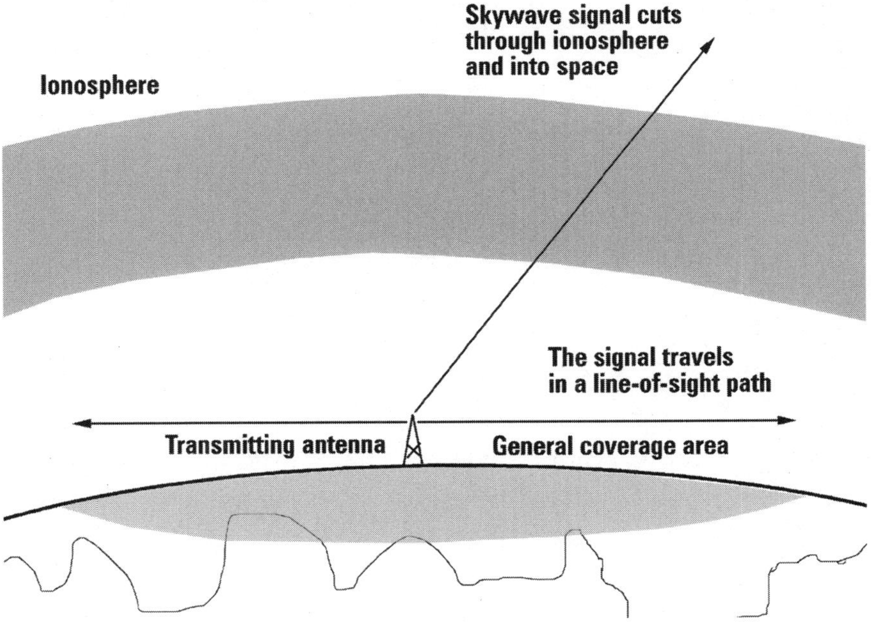

Fig. 11-4 A figure that shows the typical propagation of a TV signal.

The curvature of the earth affects the TV signals because of the basic line-of-sight path that the signals take. If the TV signal is traveling in a horizontal path that is parallel to the ground around the antenna, it will eventually draw away from the earth and cut through the atmosphere and in to outer space. The only ways to receive a distant TV signal are via odd "skipping" conditions that reflect the signal back to earth (such as ducting or sporadic E skip) or by placing an antenna thousands of feet up in the air! This is one reason why locations on top of mountains are so effective for receiving—there are no (or few) obstructions in the signal path and they elevation of the mountain above the average terrain can foil some of the curvature-of-earth effects (Fig. 11-5).

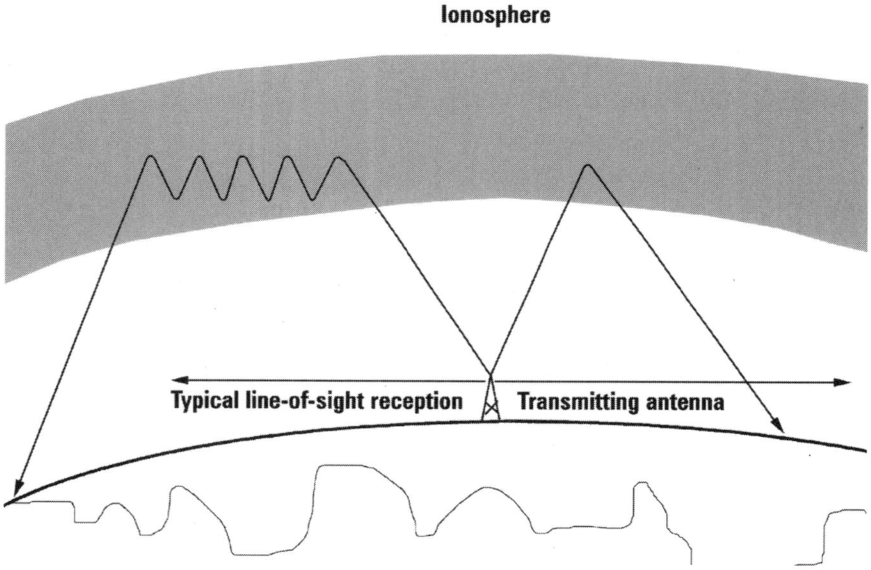

Fig. 11-5 TV propagational anomalies: examples of signals for ducting (left) and sporadic-E skip (right).

So, the only way to improve the reception in these cases is to raise the antenna to a higher location. The only effective way to permanently raise the antenna is with a mast. Tree-top locations are not preferred because the trees often get in the way of the antenna and put the antenna in danger of falling limbs or being toppled. Masts are the only way to go.

MASTS

A few different types of masts are widely available for TV antenna applications. The tripod mast, tower system, and pipe mast are all very commonly used, depending on the type of antenna used and the height at which the antenna will be mounted.

MAST TYPES

* The *tripod* mast is a three-legged mast that is typically mounted on a roof peak (Fig. 11-6). The steel tubing legs are about three to five feet long. The feet can be mounted directly to the roof or a board that is mounted on the roof, and the antenna-mounting pole slides down into the tripod. Some tripods are guyed, but if the antenna-mounting pole is short, most can be self-supporting.

Fig. 11-6 A tripod mast for mounting an antenna on a roof peak . Channel Master

Tripods are used when the antenna doesn't have to be raised to a great height above the nearby terrain.

* The *pipe mast* is another low-height form of antenna mast. This one is very simple: A long pipe is mounted alongside the building (Fig. 11-7). The antenna and rotor (Fig. 11-8) are bolted on top. Pipe masts are simple, effective, and inexpensive. Unfortunately, they are also very weak if the mast is extended too

Fig. 11-7 An antenna mounted on a pipe mast. Notice the strap around the chimney. Channel Master

far without support. If a pipe mast is extended more than about eight feet above its last mount, it is usually (for good reason) guyed. If an unguyed, long pipe mast gets caught in a wind or ice storm, it could cause roof damage, destroy the antenna, and possibly injure someone. Because guying is a real hassle, many people choose tower systems.

* The *tower system* is a self-supporting mast (although tall versions are always guyed) consisting of three parallel metal pipes with welded cross bars. These towers are available in mast sections that are about 10 feet long. Each of the sections fits together and you can stack a number of these sections to produce a really tall tower, although it has to be guyed if it's unsupported for more than about 40 feet (check the manufacturer/installer information if you plan to install one).

The big problem with the tower system is cost. You will need to spend quite a bit on materials, including sinking the bottom section into a big chunk of concrete in the ground. Then, you will

Fig. 11-8 The rotor control (left) stays near the TV and is used to control where the antenna is turned. The rotor (right) mounts between the antenna and the mast and it performs the actual work of turning the antenna. Channel Master

have to pay a company to deliver all of the heavy sections and install the thing.

FINDING AN OLD MAST

Television antennas have long been supported by aluminum, and many antennas installed 30 years ago are still supported on their original masts. Your local television shop or hardware store can probably supply you with an aluminum mast. Some radio and television antenna masts are commercially available for prices that range from reasonable to expensive. If you want to find the least expensive mast available (including those that you could build yourself), you might consider checking the weekly "shopper" papers for used television masts. Now that cable television has conquered off-air television in the cities and large towns, old television antennas and masts can be picked up for almost nothing. Before you agree to take a mast, you should probably be sure that it has already been taken off the roof or has been disassembled. Otherwise, you could damage (or be blamed for damage) the owner's roof or the house itself.

A good rule of thumb is to use the standard television antenna mast materials only for lightweight antenna systems. The medium- and heavy-duty antenna mast materials can be used for larger antennas. Proper guying techniques must be used for aluminum masts. If a section is improperly guyed, it will bend and you will be forced to replace it. Never use a bent section—even if it has been straightened. Bent sections will be much weaker than the rest of the mast. The bent section will eventually bend and/or break. Then, it will have to be replaced, costing a lot of time and money because the whole antenna and support system have to come down. Galvanized heavy-duty tubing in slip-together pieces is available at most hardware stores and it makes an excellent permanent and portable mast material. It is strong enough to support the largest antenna systems and it is not very expensive.

If you purchase a commercial aluminum mast or tower, be sure to follow the directions for installing it. If you find a good deal on a used aluminum mast or tower, it's best to check with a local television shop to see if they have directions for assembly/disassembly of that particular type of antenna mast or tower. If you can find the instructions, it will probably save you a considerable amount of time and energy.

Few people would want to build their own aluminum antenna mast because used masts are less expensive than building your own mast. Aluminum welding is much beyond the scope of this book, so if you are a welder and decide to take that route, be sure to follow safe antenna construction and installation practices.

GUYING

For most lightweight masts up to 45 feet high, stranded steel wire is preferred for guy wires. Nylon rope is used to guy some homemade masts, but that's because those masts are short and lightweight.

Steel wire is carried by most hardware stores and is usually recommended for antenna mast erection purposes. Several lengths of this cable can be overlapped and run in parallel for added strength when used for masts that will need to withstand high wind conditions or are higher than 45 feet.

Under certain operating conditions, guy wires can act as an antenna. This is not usually an advantageous situation because it will often detract from the performance of the main antenna system. Strain insulators are usually adequate for most frequencies. The only type of insulator that should be considered for guying purposes is the *ceramic egg* variety. This type of strain insulator will not allow the guy wire to separate—even if the ceramic insulating material shatters. The two sections of guy wiring are insulated from each other, but are intertwined within the egg insulator (Fig. 11-9).

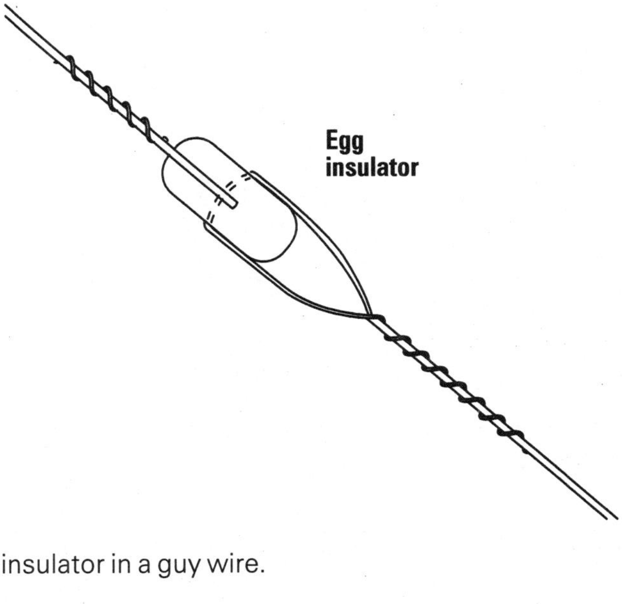

Fig. 11-9 An egg insulator in a guy wire.

Anchoring guy wires is best accomplished by connecting them to pipes (driven into the ground) or to "dead man" supports that have been buried into the ground. For best results using pipes, bend the last few inches of the pipe over (at approximately a 45° angle). Then, pound them into the ground (with the bent side facing away from the antenna) with a sledgehammer. Having the end of the pipe bent will help keep the guy wire from pulling off the end of the support. Also, if you pound the ends of the pipes to the ground, it will prevent those who walk through the yard from getting nasty scrapes and bruises on their ankles. As many times as I have tripped on tent stakes, this could be a real advantage (Fig. 11-10).

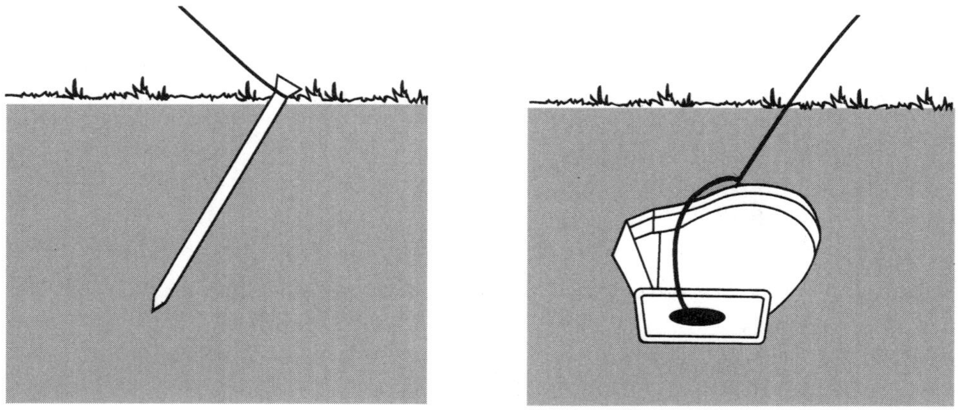

Fig. 11-10 Holding down those guy wires: using a stake (left) and using a buried toilet as a dead man (right).

It is possible to guy masts to a convenient tree or building, but you must carefully choose and install both. If you improperly choose or install your guy wires to one or more of these objects, you could damage or uproot a tree or create some havoc at your house. For small masts that support very light antenna systems, a 4-foot length of pipe driven into the ground at a pronounced angle will usually provide adequate anchoring. Choose your materials carefully when building these anchors.

The "dead man" support is merely a large piece of metal or concrete that has been buried about 3 or 4 feet deep. Some objects that make great dead man supports are large car parts (such as wheels, bumpers, engine blocks, etc.), old toilets, and concrete sidewalk slabs. One of the great aspects about using large junk to support your antenna masts is that you can get rid of this stuff for free and it can serve a practical purpose (as opposed to sitting for decades or centuries in a landfill).

Faulty pipes, used to anchor guy wires, have been known to fly out of the ground at lethal velocities when a large gust of wind provided a sudden pressure. All too often, the anxious builder will simply drive a few tent stakes into the ground to anchor the mast and antenna system that he or she has spent days building, only to find the entire system lying on the ground the next day. All materials, no matter how small, play an important role in the construction of your antenna and antenna support systems.

CONCLUSION

Most people who are serious about their television viewing opt for DBS or cable TV. Still, if you are interested in improving your reception, yet don't want to pay thousands of dollars per year for satellite or cable TV service, improving your standard TV antenna is the way to go. For more information on DBS/satellite TV, see Chapter 10.

12 Troubleshooting

These days of troubleshooting and repairing your system are significantly different from 30 years ago. As mentioned in the next section, it is very difficult to actually repair any modern electronics equipment at the component level these days. Instead, either the problem is tracked down to an entire board, which is replaced, or to a moving mechanical assembly, which is replaced or repaired. On the other hand, the functions and equipment are currently more complicated than ever before. As a result, much of what can go wrong with your system actually occurs on the *outside* of your equipment boxes, rather than on the inside.

Although this allows many more potential problems to occur, the good news is that you can troubleshoot and fix system-stopping problems without having an electronics background, just some patience.

Wiring problems

One of the most common types of problems in today's home theater systems (and most any electronics, in general) is simply in the wiring. If the cables aren't connected properly, the system won't work. It's that simple.

And it's just as easy to connect a system together, but many people are intimated by the looming home theater cabinet and the ominous dark corners, laden with dusty knots of audio-cable spaghetti. To many of the people reading this book (who bought it because it was in the electronics section at a bookstore or because they have years of kit-building and repairing experience), this might seem a bit silly. But, I have been asked numerous times by friends and coworkers to trace down audio- and video-cabling problems. The most important fact of all is that plenty of good equipment has been thrown away because it wasn't connected properly or because a cable malfunctioned. Really. I have electronics-scavenging friends who have discovered perfectly good equipment, thrown out by individuals and businesses alike, that had one bad cable.

Every wire or cabling problem can be brought down to one of two different troubles.

* The cables aren't plugged in properly.
* One or more cables are defective.

If you have a problem with your system, first use logic to follow your way through the cabling and switches. For example, the problem might be that no video is present. First, check that the TV actually turns on and the screen comes on. If not, immediately check that it's plugged in and that the surge protector is on. If the surge-protector light doesn't turn on, try the other outlet, then the circuit breakers, if necessary.

If the TV does turn on and goes to a blue screen, try playing back a number of different video sources. For example, if you can receive cable TV and DVDs play fine, but there's no VCR video, you've just narrowed down the problem. In this case, you would check to see that the VCR is playing back OK. If not, it's a VCR problem. Next, be sure that the TV is on the proper channel for playback from your VCR. If the playback is working fine, chances are good that the problem is located between the VCR and the A/V amplifier.

First, check the buttons on the front of the A/V receiver (or on the remote control) to be sure that the video is set on VCR. Then, check all of the cables from the VCR to the A/V amplifier. One common problem is that cables are mixed up--the *in* cables are switched with the *out* cables (and vice versa) or the *audio* and *video* cables are reversed. Sometimes the problem is that people just aren't careful when connecting the equipment. At other times, the cables are plugged in while the backside of the equipment is hard to see and difficult to access. It's easy to make such mistakes under these conditions.

By now, the problem should have been resolved because the remaining possibilities are much less likely to occur. If problems still occur, there might be a mechanical failure in one of the cables. Cables fail regularly in applications where the cables are frequently bent or twisted; however, most people just buy their equipment, set it up, plug in the cables, and let everything sit. Not much of a chance for the cables to get damaged there. Still, if a cable has been used previously (such as at some of the used hamfest specials that I've mentioned at times throughout the book), it might have failed.

CABLE AND CONNECTOR FAILURES

The two types of cable failures are *opens* and *shorts*. In an open, the wire breaks and the two pieces are no longer in contact with each other. In a short circuit, the insulation breaks down between the wires and they "ground out." With an open, the signal will disappear; with a short, a "grounding buzz" appears. In either case, the wires are loose and will occasionally touch together and pull apart, creating a "touchy" condition where the wires sometimes work and sometimes don't. These problems are collectively called *intermittents* because of their intermittent operation. Intermittent audio or video cables should always be replaced because the wires have been broken in a high-wear area and they are sure to cause future problems (Fig. 12-1).

A related problem occurs with the cable lead-ins for cable TV (especially) and satellite (analog or DBS) and TV antennas. Very often the cables are not manufactured as a complete unit, but the connectors are crimped on the ends. After years of use, it is common for the connector to pull out from the cable itself. When the cable starts to pull out, the signal still jumps the gap and makes it to the connector, although at a weaker signal level, causing the picture to appear snowy. If the picture appears snowy on all channels, this

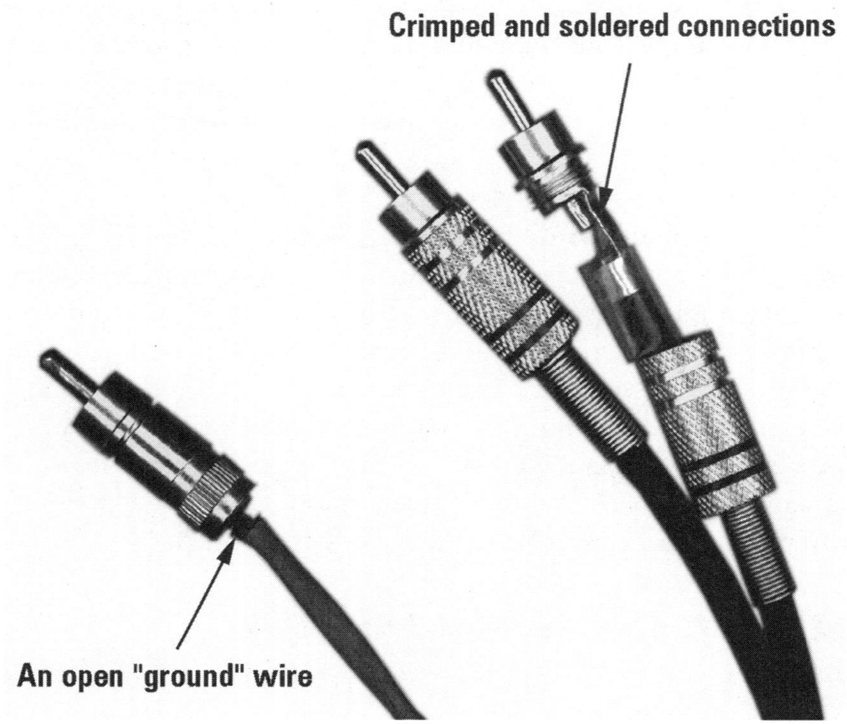

Fig. 12-1 Two sets of cables. At the left, a well-used cable with a connector that started to pull apart from the cable; then the "ground" wire broke. At the right, a good cable, with one connector sleeve unscrewed, showing the crimped and soldered connections within.

connector is one of the first places to check. If you do discover that the connector is loose, push it back on the cable and crimp it (if you have a crimping tool) or wrap a piece of electrical tape around the connector and the cable to hold it in place.

REMOTE-CONTROL PROBLEMS

One potential external equipment difficulty is the remote control. If erratic behavior or total failure occurs, you don't have to pull out the test equipment and an engineering degree to repair it. The most common problem is, of course, dead batteries. If the remote control dies, always replace the batteries first.

Another possibility is that the infrared transmitting LED or sensor is blocked or damaged. It could be blocked by dirt or a build-up of hand grease, rather unpleasant to consider, but true. If so, just get a clean rag or paper towel and clean it.

If the remote still seems dead, take a look around the battery terminals. After months or years of abuse, the wires that connect the battery terminals to the rest of the electronics could break off. Check to see that the wires are connected and that the batteries are solidly against the contacts. Also, check that you have the battery polarity correct (Fig. 12-2).

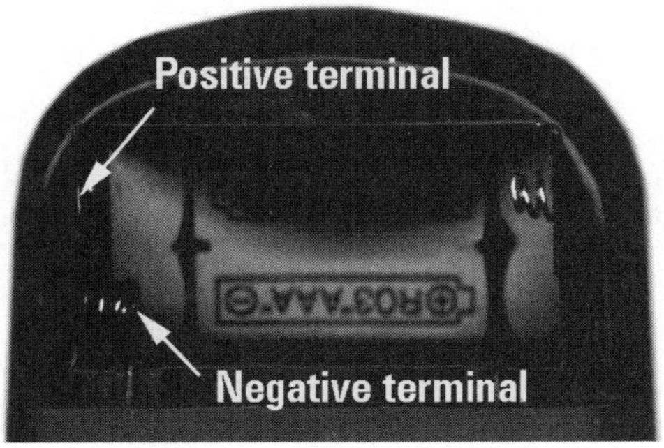

Fig. 12-2 The inside of a remote control's battery compartment showing the battery terminals. Any opens would occur on the inside of this plastic shell.

If you lost the remote control for a few months and the batteries sat idle, there's a chance that they leaked a corrosive liquid inside of the remote control's battery compartment. After this happens, the liquid typically corrodes the terminals, leaving them rusty with a white, powdery residue. You must either replace the contacts or the remote control after a case of battery corrosion.

LEDs rarely need to be replaced, but remote controls take a lot of abuse and the LED is exposed. If nothing is working, you can see that the LED isn't lighting, and you would toss the remote in the trash otherwise, try replacing the LED. Remember two things when purchasing a replacement: Buy a component that will fit the hole in the remote control and buy an infrared LED (IR LED). For tips on soldering and safety, see the following sections.

TAKING A LOOK AT THE EQUIPMENT

From an electronics repair point of view, it is both good and bad that the 1950s have passed. In those days, huge, well-built components were screwed and bolted into large steel cabinets. The parts were built well enough that they were expensive, but didn't break down frequently. And they were large enough that they could be replaced without difficulty. Believe it or not, many of the TVs sold in the 1940s and very early 1950s were actually electronics kits that you could build yourself!

Today's advantages are much less expensive components and miniaturization—tiny components are crammed so tight that it seems that only a malnourished seven-year old with tweezers could have assembled the circuit boards. The disadvantage is, in turn, that only that same seven-year old could disassemble the board without destroying everything else in the process. As a result, in almost all levels of modern electronics, actual repairs are no longer made; instead, the malfunctioning circuit board is pulled out and replaced with another board.

There is very little that you can do to repair failures on these boards, except purchase a few junker units that have the same model number as the unit that you own, isolate the problem, and interchange the failed board with one that will work. This should be a great possibility, but chances are that you will not find a number of broken-down units that are exactly the same as yours. If you do find a number of broken units like yours, chances are that you picked a poor model and you might be better off dumping it than trying to repair a cheaply designed piece of electronics. Also, when a breakdown occurs, the chances are good that it will occur in the same parts most of the time. This is the weak link. One of my friends has a stack of nonfunctional radios of the same model and he has said that all have either one of two different parts failures. Fortunately, my friend knows that particular model well enough to fix these failures quickly and easily.

Most audio and video components are built using the board-for-board method, except for dedicated power amplifiers. Dedicated power amplifiers, especially the high-quality American-made models, usually use large, heavy components soldered onto thick circuit boards. These amplifiers could be repaired, but they often don't break down unless they have been mistreated. Even then, it is usually the power transistors (usually field-effect transistors) that go through meltdown. If your amplifier contains vacuum tubes, they will slowly wear out, through no fault of your own. But, unless you use the A/V amplifier as a switch box/preamplifier and feed the outputs into a dedicated amplifier, you won't be using one.

If you can con the manufacturer out of a repair manual, then you will have a much better opportunity to repair your equipment. However, you probably will not be so lucky. Unfortunately, most companies only distribute this information to their authorized dealers, which forces nearly everyone to go to these dealers for repairs...or buy a new unit. In the 1950s and 1960s, a number of American electronics companies provided extensive support for their products with descriptive lists of components, test point information, troubleshooting information and tips, etc. These days, everything has changed and you are lucky if you can get any free information about any electronic gear.

This chapter is primarily intended for people who have anywhere from no experience to intermediate experience with electronics in general or A/V systems in particular. It is not a technician-level book; i.e., it's for consumers, not electronics enthusiasts or repair technicians. You won't be able to use this chapter to discover that voltage regulator Q74 is blown in your particular TV and that you should replace it. Information of this nature has filled a number of large repair books on its own. Instead, this chapter covers a few of the troubleshooting and repairing basics that are relatively easy for the beginner to perform and that don't require the use of expensive test equipment. Also, because modern video equipment is so tightly packed and is often built to discourage repairs, it is best to be very cautious when approaching the repair of a unit.

TRACKING DOWN THE PROBLEM

Even if you have no intentions of repairing your own equipment, it is important to search out the problem and tell the repair technician where the problem is and what is occurring. This will help the technician find the problem in less time and it could save you some money.

In a few of my books, I have mentioned seeing an episode of "60 Minutes" several years ago with a segment that dealt with the integrity of various electronics repair shopkeepers. In it, they had a woman take a VCR with a simple problem (I believe that the drive belt might have been off) into different electronics repair shops in one city. Several of the parts were marked with an ultraviolet pen. The result of the study was that most of the shops were not honest. Some claimed that parts were replaced that weren't; others claimed that they had performed repairs that hadn't been done, etc.

One of my friends read this story and noted that his family had experienced a similar situation when he was younger and his parents felt less confident in his TV-repair capabilities. In this case, he knew what was wrong, but they took the equipment to the shop, which recommended an expensive, unnecessary repair. From that point on, the parents let their son repair any of their malfunctioning electronics equipment! This isn't to say that all electronics repair shops are dishonest—far from the case—but you should always be as informed and knowledgeable as possible on the subject. Nothing will deter getting ripped off on a repair like knowing what has failed where and being able to talk about it with a repair technician.

You should always first check the wiring around where you think the problem is (and possibly over the whole system, if you fail to locate the trouble). Faulty wiring or connections will void the entire troubleshooting process if you don't immediately check these possibilities.

If, for example, your system does not work and no video is reaching your TV from a VCR, DVD, or laserdisc player, first check the buttons and system programming. Maybe the VCR button is pushed in, but the DVD is playing. Or if everything works, but there is no audio, first be sure that the Mute button isn't pushed in.

After you have checked everything, look over your system wiring. Are the units turned on? Is your system plugged into the wall outlets? If you have equipment that plugs into an amplifier, try pulling the plugs out and plugging into the wall outlets instead. Or maybe the system lights up, but you aren't getting any sound out. Have you recently added any components that could have created problems with your system? If you did, check the component's external wiring. Next, check for common sense things that you might have missed. Did you forget to plug in any audio cables when installing or rearranging the system?

If these problems aren't applicable, check the displays and power-on indicators on the various components. If you aren't getting any sound, does the component even turn on? If not, you should check the connections to the unit. As complicated as today's audio/video components are, it is easy to accidentally plug the input cords into the output jacks (or vice versa). An overlooked connection can cause the whole system to go down or part of it might not work.

Determining faults in a system is a matter of taking educated guesses and almost randomly substituting other components for testing purposes. The rationale is that if one part isn't working, you will be able find that problem by substituting different components one at a time until the system works. From this point, if the problem is in a complicated component, such as a DVD player or an A/V amplifier, you might not be able to narrow it down further without using a multimeter, a signal generator, and an oscilloscope, so you might be stuck at this point if you don't have this equipment.

Of course, these tracing examples could be drawn out for 101 different minor problems. Rather than delve into all of these problems, which you probably won't ever encounter, and bore yourself to tears in the process, this book covers a few basic problems. If you have not had much experience with repairing or troubleshooting electronics equipment, this simplified approach should be more helpful than a deep, all-inclusive look at the topic.

ORGANIZING YOUR WORK

The procedures to disassemble audio/video electronic equipment are very complicated and you must be very careful or you might wind up with a few "extra parts." Before you take anything apart, be sure that you will not be voiding the warranty of the product. To prevent tampering, some companies even place "Void if torn" stickers along the edges of the chassis. If your product is still under warranty, don't mess with it. Just send it back and save yourself the extra money. If the component is past warranty and has these stickers, don't worry about tearing them and cautiously looking around for the problem.

As you disassemble the electronic components, don't just dump out parts in a pile. Write down the parts that you have removed and either put them in labeled sandwich bags or in small containers. Better yet, if you have a Polaroid camera or a digital camera, take a photo or two before you start the disassembly process. I have torn apart different components many times and felt that I would know for sure how to put the parts back together. Before long, it seemed that I had the entire project into hundreds of little pieces, and I only remembered how to fix half of the parts in the proper location. You should even draw a diagram of the assembly/disassembly procedures (even if you have a photo) so that you can prevent your projects from remaining hopelessly disassembled.

SAFETY PROCEDURES

Safety procedures are never fun because they slow down your progress. I guess I'm just too impatient for safety rules, but it seems as though I have to screw up several times severely before I get the hint that I should be more careful. Then, after a few traumatic experiences, my techniques improve. Fortunately, I've never come close to electrocuting myself, but I have grabbed the soldering iron by the wrong end a few times! Believe me, you're better off avoiding stupid mistakes by reading about them than by doing them yourself!

If you want to learn about repairing equipment, but you are not a technician or you have very little or no experience, read some books that are dedicated to the topic and buy some inexpensive nonfunctioning equipment to experiment with. One of the worst things you could do is to try to repair an expensive piece of equipment (such as a $1900 TV) with minor problems, cause more damage to it, and not have enough money to replace it.

If you have had some experience working on electronics gear and feel comfortable about tinkering, be careful and connect the power only when taking measurements at the manufacturer's listed test points. When testing and taking measurements on a unit, it is much better to take it out of the entertainment center. Go to your workbench. This way, you have work space and sufficient light, rather than trying to solder upside down on the floor with molten solder and flux splattering on your carpet and/or the rest of your equipment. It is best if you have an open work area or an electronics workbench that is clean, stocked with parts, and well-lit.

ELECTROCUTION AND ELECTRICAL SHOCKS

The one serious drawback to troubleshooting and repairing electronic equipment, aside from the possibility of damaging it beyond repair, is the possibility of electrocution. Sure, you work with the voltage from the wall every day, so it might seem relatively tame. It's not. One buzz from the wall could be your last! More than enough to kill you, 110 volts and the corresponding 60-Hz cycles (in North America) are said to be near the optimum frequency for stopping a heart. The voltages in Europe aren't any safer—220 volts at 50 Hz--so you must be excruciatingly careful with electronic equipment wherever you live.

If you make a mistake and nothing happens, don't assume that you'll always be so fortunate. For instance, you could be electrocuted by touching a "hot" wire carrying line voltage, but you might actually escape it. This doesn't mean that you should be careless when repairing electronics. Instead, you should take extra precautions to increase your chances of survival if you touch a "hot" wire. Electrocution is a real possibility if you are well grounded and if you are making a strong contact with the electricity (such as if your skin is wet, if it is entering through a cut, etc.). To prevent being grounded, don't stand in water, wear rubber-soled athletic shoes, and stand on a rubber mat, if possible. Electricity always takes the shortest path with the least resistance to ground. So, if you are standing on a mat and get shocked, you are actually a giant resistor, but if you are grounded through a puddle on the floor, you are a giant conductor. From a safety standpoint, you are much safer being a giant resistor (however, you are much safer, and happier, not being an electrical component at all!).

Because of the shortest-path-to-ground rule, it is best to avoid having electricity grounding in a path that leads through the heart (Fig. 12-3). For this reason, one commonly used rule in electronics

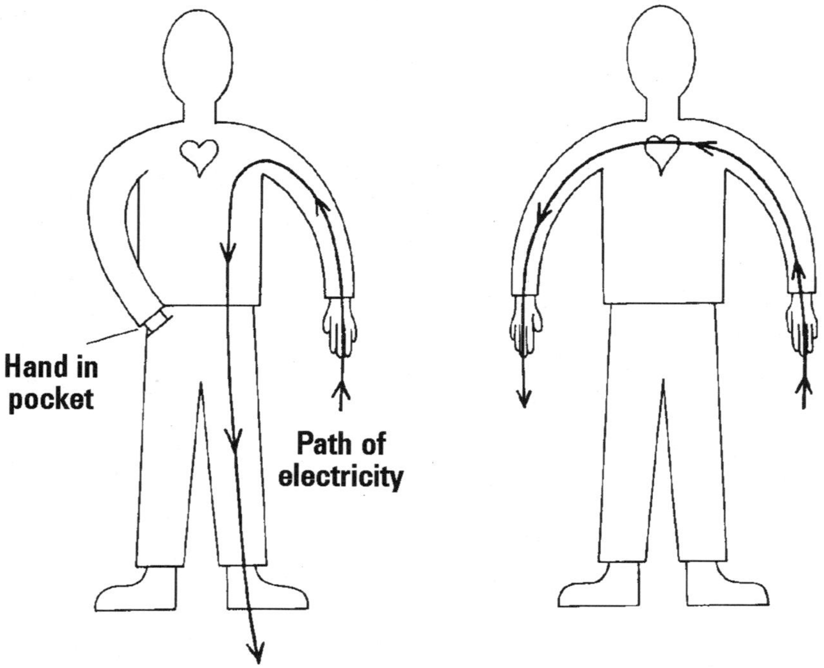

Fig. 12-3 With a hand in your pocket, the electricity will pass through your arm and out from your leg. Otherwise, it will travel through your heart. Loompanics

is to work with the left hand in a pants pocket (one of your own). Thus, if you get shocked with a great amount of power, the electricity shouldn't pass through your heart and you won't get killed. It's a simple and effective theory, except for one hangup. It's really tough to work with one hand in your pants. After years of working with your hand in your pants pocket, you probably won't get electrocuted, but your scarred hands and mangled projects will testify as to the clumsiness of the arrangement.

As stated previously, be sure to unplug the unit that you are checking (unless you are taking measurements at various test points) so that you don't cause yourself bodily harm. And remember from the operational theory of capacitors that capacitors store electricity until that charge is dissipated. This charge doesn't quickly dissipate while it sits in a circuit; you could easily dig into a TV weeks or months later and find out that the heavy-duty capacitors still pack a wallop. In video equipment, you won't find too many high-capacity capacitors, but beware of those in the power supply of any piece of electronics, and especially those in TVs and power amplifiers.

The usual method for discharging capacitors is to cross the leads with a large plastic-handle screwdriver (Fig. 12-4). However, to

Fig. 12-4 Discharging a capacitor with the old "screwdriver across the leads" method. Loompanics

prevent the possibility of having a charge jump over the shaft of the screwdriver and jolt you, it is best to attach the screwdriver shaft to ground via an alligator clip. Be sure to discharge each large capacitor several times because most capacitors will still store some electricity even after you have discharged them the first time. This method has been used at TV repair shops for decades, but some people say that it causes slight damage to the capacitors each time.

For a more scientific discharge method, cross the leads with a resistor of a value that has been recommended by the manufacturer. This method can be clumsy and annoying because you have to dig up different resistors and shove them into the guts of some equipment without bumping against other components or shocking yourself. Some technicians use a special capacitor discharger, which slowly dissipates the stored voltage without the possibility of damaging the capacitor. Check with some of the large electronics-supply mail-order companies for availability.

SAFETY WITH ELECTROSTATICS

To protect electronic equipment from electrostatic discharges (also known as *static electricity*), you must take almost the opposite mindset and approach than you did with protecting yourself against dangerous voltages. That doesn't mean that you should stand naked in a puddle of water while groping the insides of a plugged-in TV, but it is still somewhat different from protecting yourself against electrocution.

Many of today's microelectronic components are intended to handle only minuscule amounts of power. Because of the push for miniaturization in electronics, it is better for the parts to be used closer to the damage tolerances than to be forced to build the piece of audio/video equipment into a much larger enclosure (larger equipment is also more expensive). As a result, it is very easy to damage equipment with levels of electricity that we would have never considered to be harmful to anything just one or two decades ago.

However, with modern integrated circuits, even a small blast of static electricity can melt a tiny hole through the package and destroy one of the microscopic components or interconnecting wires. Blasts of static electricity will either destroy or impair the performance of the integrated circuit. Even the static electricity discharged from your body is destructive. Just wearing tennis shoes and walking across the carpet on a dry day will cause an electrostatic charge that is easily powerful enough to destroy a typical integrated circuit. According to an article in *Electronic Servicing and Technology*, one person (depending on the clothing and other environmental conditions) can build up a charge of static electricity that exceeds 35,000 volts!

As you can see, it is very difficult to handle integrated circuits without damaging or destroying them. To prevent damage, it is important to wear a grounded wrist strap (Fig. 12-5). The grounded wrist strap is a conductive strap that is connected to ground via a wire. This method is effective because the ground bleeds off the electricity (rather than allowing it to suddenly build up), then releases it through the integrated circuit that you are handling.

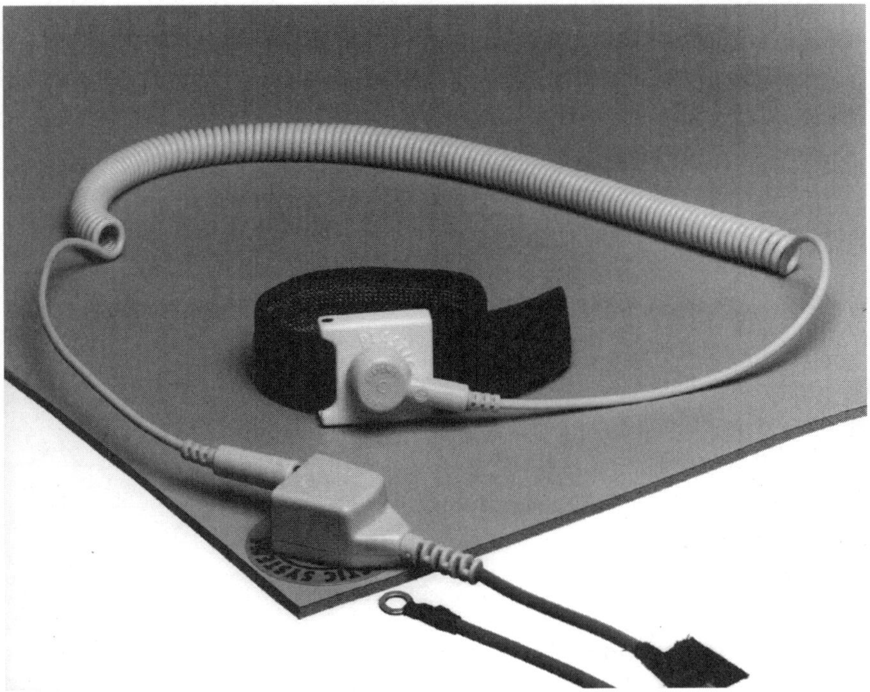

Fig. 12-5 A grounded wrist strap, which is necessary when handling modern IC electronics. Plastic Systems

So, if you are working on complicated integrated circuit-laden units, such as receiver/amplifiers, be sure to use grounded wrist straps. Otherwise, even getting too close to the unit (without touching it) could cause your body to discharge electricity through it (via arc-over) and damage components.

SOLDERING IRON SAFETY

The biggest problem that you'll likely face with soldering irons is not the intensity of the heat, but rather that they don't heat up fast enough. But they're hot enough to melt solder, what more could you want? When troubleshooting and repairing modern miniaturized electronics, it's best to use a soldering gun. Unlike the "pencil" soldering iron, the soldering gun is pistol-shaped and is activated by a trigger. These models are typically rated at between 40 and 100 watts, as opposed to 25 watts for a "pencil" soldering iron (Fig. 12-6).

The problem with using a "pencil" soldering iron is that it doesn't get hot enough to quickly melt solder. When you make a soldering connection or desolder a component with an iron, you have to hold the tip in place for a while—heating up and possibly melting the component. I burned up a few surface-mount microswitches within the past year or two using soldering "pencils." Another problem is if you decide to solder A/V cables, the soldering iron might not be hot enough to thoroughly heat the wire, producing a "cold" solder joint. With a soldering gun, you can "zap" the connection and move on. In some cases, you might want to connect a heatsink or a pair of hemostats (commonly available at radio hamfests) to the component that you're soldering so that the heat will be transmitted into the heatsink (or hemostats), not the component.

CLEANING

One important factor of electronics equipment maintenance is to keep it clean. Dirt from hands, cigarettes, or dust particles can cause buttons to become intermittent or fail altogether, or can wear down moving parts. One communications radio receiver that I have is more than 40 years old. A few years ago, a friend and I unscrewed the case and had a look inside. The circuit board underneath the chassis was so piled up with dust that the voltage actually arced over from one component, through the dust, to another component! Fortunately for me, the older tube equipment is able to withstand many abuses, and the receiver continued to work fine. Modern home theater equipment would not be so lucky.

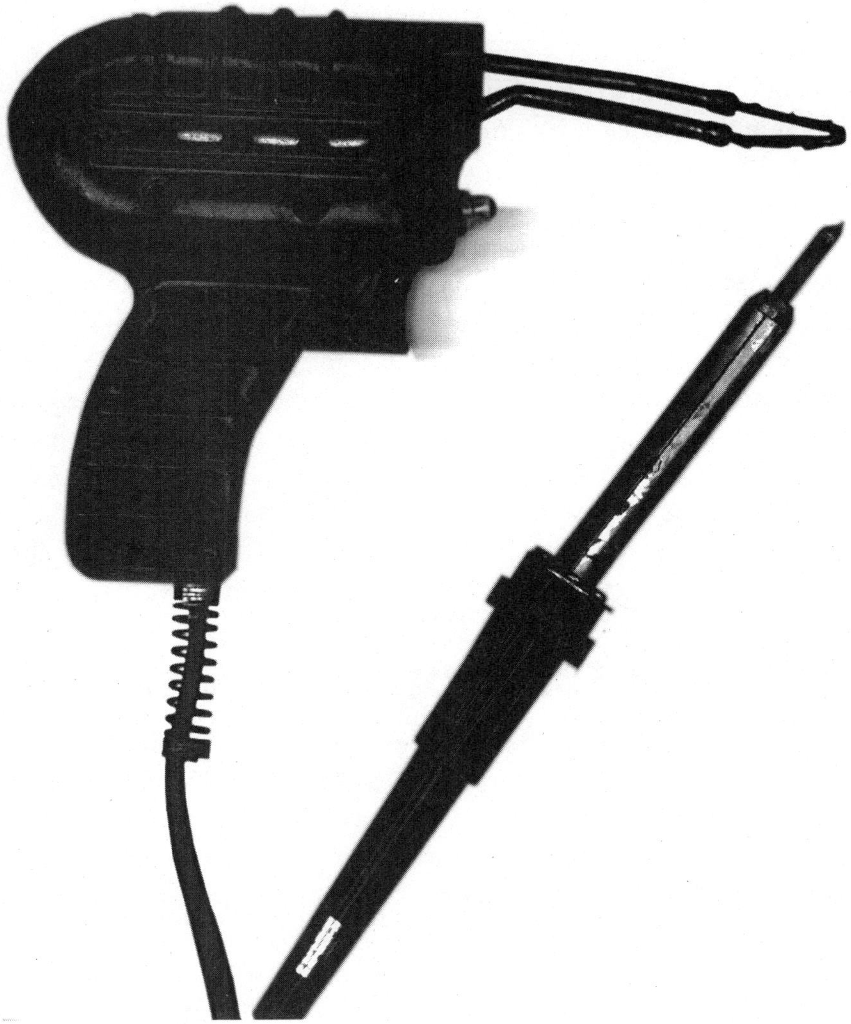

Fig. 12-6 A soldering gun (upper left) and a soldering pencil (lower right). Notice the burned, disfigured tip on the soldering pencil.

If you regularly repair modern electronics equipment of any type, you should always keep a can of compressed air nearby. Just after you remove the screws and the outside of the cabinet, you should have the can of compressed air handy to blow out any loose dust particles or "dust bunnies." To avoid problems with static electricity (as mentioned in the previous section), try to avoid touching the boards or parts with your hands. If the equipment is in really bad shape, you might want to try rubbing alcohol and cotton swabs. For example, one VCR that I worked with had dirty, water-stained PC boards and needles from a hemlock tree covered the insides! In this case, I opened the cabinet, turned the VCR upside-down to remove most of the needles, and I used rubbing-alcohol-dipped cotton swabs to clean the parts and the boards.

There aren't too many different practical ways to reduce airborne dirt and dust particles in your house. If you currently smoke, stopping would help. Forced-air heat compounds the dust problem, but switching to electric or hot-water (via radiators) heat just to reduce the wear on your electronic equipment is a bit extreme. If you feel that airborne dust particles are a problem, you might consider purchasing an air purifier for your home theater room. One of these units (available under several different names) sucks in air from the room and filters out the dirt and dust.

DVD AND LASERDISC PLAYERS

Like all modern audio electronics (except power amplifiers), DVD and laserdisc players are very complicated units that are packed with tiny components, some of which are sensitive to static electricity. Thus, these units are very difficult to repair, but it is fairly easy to perform some easy maintenance tasks that will keep your discs running properly. The troubleshooting information included in this book is limited. For some related information, read *Troubleshooting & Repairing Compact Disc Players (3rd Edition)* by Homer L. Davidson. Currently, there are no books about troubleshooting and repairing DVD or laserdisc players, but CD players are very similar in operation, so this book should be a significant help.

SAFETY

Laserdisc, DVD, and compact disc players of all sorts can be very dangerous to work on, both because of your own safety and the delicacy of the player itself. They operate somewhat like a record player, but rather than using a needle to pick up vibrations from the disc, a laser beam is used to read the underside of the disc. You can disable the mechanism that turns off the laser and put your hand in front of the beam for a long time. It won't suddenly cut through your hand, and you won't feel any pain. For that matter, you won't feel anything. Despite the first impression, this laser is not totally harmless. It can severely cause damage to your eyes if you look at it. So, don't! And to keep your eyes safe when working on a powered-on disc player, place a disc in the machine if you need to poke around inside. The disc will cover the beam. Obviously, your eyes are very important, so be careful.

In addition to the hidden potential for physical bodily harm while working on laser-beam equipment, these devices also present a potential for self-destruction. As mentioned earlier, these players are more prone to destruction from static discharges from your body. Be sure to use a grounded wrist strap and ground your soldering iron, too, to prevent dangerous levels of static electricity from damaging the fragile integrated circuits. Some references also state that you should use a grounded electrically conductive mat under yourself and another under the compact disc player that you are working on.

Optics

Some of the components that are most prone to failure in a DVD, laserdisc, or CD player are the optics and the disc-placement assembly (my generic term for the door, drawer, etc.--don't repeat this phrase to the repair people or they will look at you for a moment dumbfoundedly). The optics are prone to failure because they easily become dirty or scratched from wear or the user's botched attempts at cleaning. The disc-placement assembly is susceptible to damage because it contains some of the only moving parts in the entire piece of equipment. The optics typically don't become noticeably dirty within several years of use. So, dirty optics should not yet be a problem with DVD or laserdisc players. And because of the nature of the optics, don't tear apart your player every few weeks to clean the unit!

Cleaning the Laser Optics Assembly

One of the most likely causes of skipping is a dirty or dusty laser optics assembly. Little effort is required to clean the laser optics assembly; it is very similar to cleaning the heads in a cassette deck. Just get some cotton swabs and a bottle of rubbing alcohol. Dip one of the swabs in alcohol and rub it over the glass laser optics assembly. If you see other wads of dust inside the player, clean those up, too, or use a compact disc player cleaning kit. Otherwise, the dust particles might work themselves over toward the laser optics assembly and cause more skipping later on.

CHECK THE SOURCE

Although chances are that any skipping that your CD player is experiencing is caused by the player itself, it's worth checking the CDs. If your reference compact disc or discs have been scratched or are smudged or dirty, they might be the reason why your player is skipping. I have heard some people claim that if the discs are scratched, they rub nail polish remover over the scratches. I haven't scratched any of my discs bad enough to cause skipping, so I have not been forced to try this method and verify that it works as well as some have claimed. However, if you feel uncomfortable using varnish remover on your precious video, you might want to check out the CD Finishing System from Allsop, which is a disc scratch repair kit. This is the only kit of this type that I have seen on the market, so if you can't find it in a store, you can try writing for information directly from Allsop, P.O. Box 23, Bellingham, WA 98227 USA.

If the discs are dirty, clean them with a plastic-safe, nonabrasive cleaner, such as a common glass cleaner. Or try one of the compact disc and laser optics cleaners that are available from a number of different companies (Fig. 12-7).

TECHNICAL ADJUSTMENTS

As stated previously in this chapter, technical adjustments are well beyond the scope of this book, especially for DVD, laserdisc, and CD players. For most work on these players, an oscilloscope and other electronic test equipment are required. There is very little that a beginner is able to maintain or repair on these units. If you run into internal technical problems, it's best to take the unit to a professional for servicing.

Fig. 12-7 A laser cleaning kit. Allsop

VCRs

VCRs often require more maintenance and repairs than any other piece of video equipment (except maybe TVs) because they have more moving parts and have to make contact with dirty materials. As a result, it is common for a VCR to croak: i.e., begin sounding bassy, altering the picture, slowing down, or eating tapes after a few years of use. Then, you question why you ever thought the VCR was so great and start looking for a new model. Then, the deck is sold off cheap or is simply thrown away. Many problems with VCRs are relatively minor and inexpensive to remedy, so it's a real shame when a deck gets pitched. Again, because of limitations in the scope and size of this book, it is somewhat limited in its troubleshooting and repairing information. For more than 400 pages worth of information on the subject, read *Troubleshooting & Repairing Audio & Video Cassette Players & Recorders* by Homer L. Davidson, *Troubleshooting & Repairing VCRs (3rd Edition)* by Gordon McComb, or *Maintaining and Repairing VCRs* by Robert Goodman.

Muffled audio/tapes are eaten

The most common problems with VCRs are that the audio is

muffled, the video starts to look less crisp, or that the tapes get eaten by the playback mechanism. There's nothing worse than having your brand new $24.99 videotape of "The Angry McCandlesses Live" eaten—especially when these problems have been occurring with increasing regularity over the past few months. It's time to end this violence before your VCR consumes your video collection!

Although muffled audio and problems with the playback mechanism seem thoroughly unrelated, they can be both different symptoms of the same problem. Of course, all VHS videotapes consist of some mixture of a magnetic substance (usually ferric oxide, chromium dioxide, or metal oxide) that has been very thinly coated onto a synthetic tape. After some wear, these oxides will eventually build up a light coating of residue across the parts that rub against the tape. The residue resembles some form of dirt, and it is exactly that: foreign matter that must be removed to preserve the most accurate sound and video reproduction possible (Fig. 12-8).

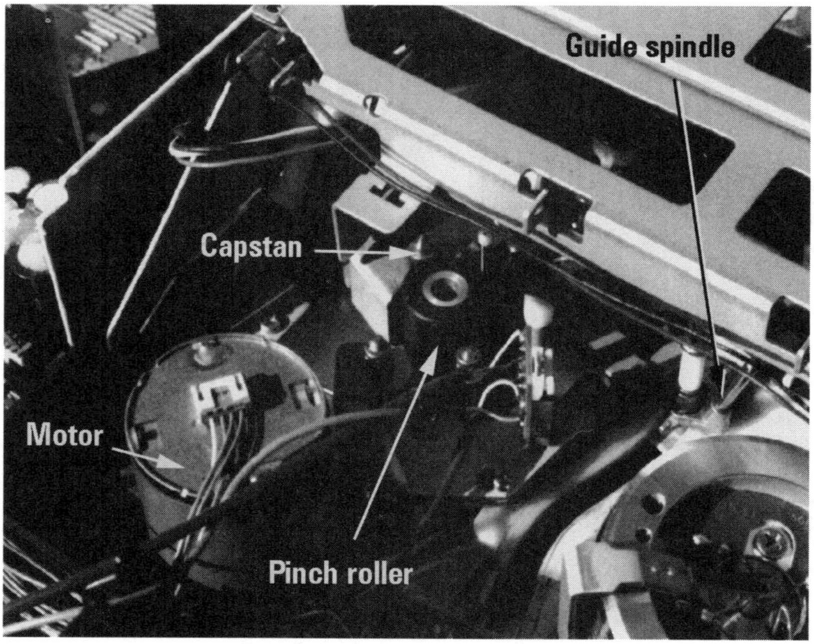

Fig. 12-8 Some of the parts of a VCR that wear the most against the tape. The pinch roller is especially notorious for tape residue buildup.

The best way to take care of these problems is to clean everything that is in the path of the tape with rubbing alcohol. With a VCR, this is not a terribly easy task, certainly not as easy as cleaning the heads of an audio-cassette deck. Cassette decks are manufactured with the understanding that you will need to clean the heads regularly; the tape door is typically manufactured to either pop off or to unscrew for easier access. By contrast, you need to unscrew the cabinet and pull off either the top or bottom cover just to get to the heads on a VCR. Then, you need to find the heads; unlike cassette decks (which have the heads right in the front), the heads are buried in the guts of the machine. Fortunately, however, the heads are easy enough to find—they're a big chrome cylinder that should be somewhere in the middle of the machine.

Just dip a cotton swab or a cleaning stick into rubbing alcohol, and carefully clean the erase head, the head drum, the pinch roller, and the capstan. If you don't know what the head drum, capstan, and pinch roller look like. All of these parts look quite similar between units that have been made by various manufacturers.

Cleaning the tape path in a VCR is much more complicated than cleaning that of a cassette deck. It's no more difficult to clean the VCR, but there are about twice as many parts for you to rub down. The capstans, spindles, and head drum should be easy to clean—just rub them with the dipped cleaning stick a few times. The pinch roller is more difficult because it is rubber and you must put some pressure against it to remove the oxide residue. After the VCR has been used for some time without being cleaned, you will notice that there is a stripe of tape residue running through the middle of the pinch roller. When you clean the pinch roller, you must turn it as you rub it clean. Otherwise, only one spot of the pinch roller will be clean. By the time you are finished, you should barely be able to see any of the residue stripe. Cleaning the heads is a common part of audio/video maintenance, and it is recommended that you clean your heads about every six months or year (you might want to clean the heads every three or four months if you record a lot of programs), although few people adhere to this advice. A clean playback mechanism is essential for the deck to perform up to its capabilities.

I recently talked to one of my friends about the wonders and difficulties of cleaning the playback mechanism. He said "Why don't you just use a head cleaner cassette?" Some head cleaner cassettes use abrasive tape to clean the heads. These abrasives can wear out the heads if you play them constantly. Frankly, after paying several hundred dollars for a new VCR, I would much rather pull out the rubbing alcohol and the cotton swabs than risk the life of that precious magnetic head. Also, some head cleaner cassettes only clean the heads, not the capstan, the pinch roller, or the other parts along the way, all of which really should be cleaned regularly. If you are looking for a cleaner cassette, be sure that it is nonabrasive and that it cleans the heads, capstan, pinch roller, and the rest of the parts in the tape path (Fig. 12-9).

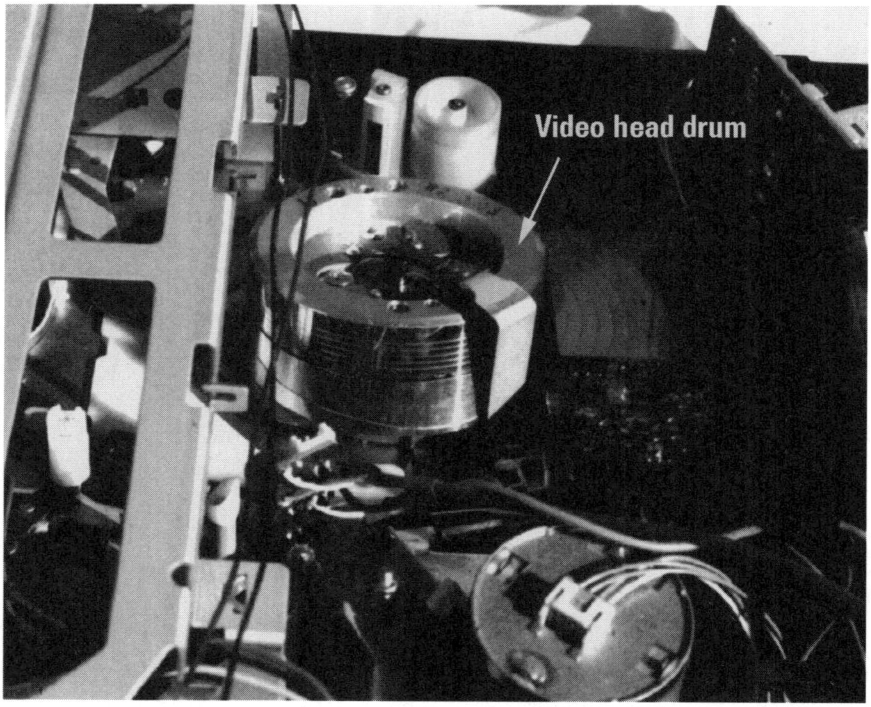

Fig. 12-9 The video head drum in a VHS VCR.

If you clean the head drum, but a few stubborn dirt spots remain, you can try rubbing across the head drum very gently with a soft, clean pencil eraser. This is primarily a last-resort type of situation: only try using an eraser if nothing else works. The heads are too fragile and costly to take chances with. Don't use older or dirty erasers because they can leave difficult-to-remove smudges. Don't use pen erasers because they can scratch the fragile tape head.

If you mostly just had problems with tapes being eaten, also look around the capstan for any remaining pieces of tape. These pieces can cause further problems with VCRs getting the munchies. Little pieces of tape are typically easy to remove from the capstans of VCRs because you can easily reach inside and pull them out. Fortunately, VCRs don't eat tapes nearly as often as cassette decks do because the tension is always on both the take-up and supply reels.

VCR turns on, but nothing happens

One of the more common malfunctions with VCRs is that the drive belts eventually slip, stretch and fall off, or break. The drive belts are fine black rubber belts that look much like thin rubber bands. They drive the various moving parts in the VCR, such as the capstan. Because of their materials and the constant wear, they will stop functioning properly at some point (Fig. 12-10).

The only recourse is to open up the case, and replace the belts. First, you have to try to find the appropriate belt for the deck. Check the owner's manual to see if a part number is listed for the problematic belt. Chances are that it won't be listed, so you will have to look elsewhere. Look up the telephone number for the company that manufactured your VCR. After a few telephone calls, you should find someone who can tell you the dealers that are authorized to distribute the parts for their equipment. Then call this company to see if they can either provide the part or at least give you the part number. Otherwise, check with some of the electronics mail-order companies that handle parts for various types of audio/video equipment (such as Parts Express--see the Appendix). Even if the exact part number is not available for your particular model, the salespeople might be able to provide a

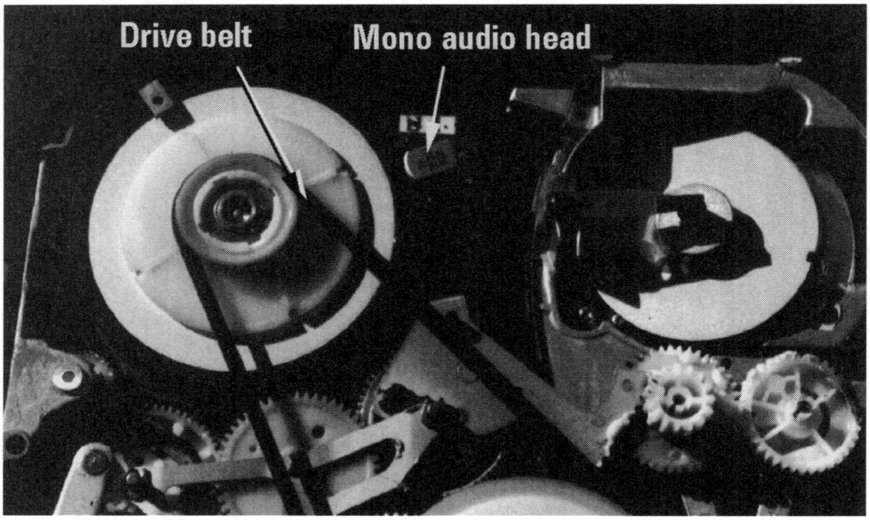

Fig. 12-10 All motion in a VCR is controlled by the drive belt.

replacement belt that has the same dimensions. If you have looked everywhere, but you still can't find a belt with the right dimensions, you could try buying a smaller belt and stretching it enough that it will fit and run properly. It's not the best solution, but it beats buying a new VCR.

Unfortunately, you might have some problems just accessing the drive belts, let alone being able to test their tension at your leisure. Before you dig around too much for addresses and phone numbers, look to see how accessible the belts are. In some cases, you cannot simply place a new belt on the pulleys; you must take apart various assemblies, panels, and boards just to reach the belt. If it appears that you cannot easily place the belt on, take the VCR to a repair shop and explain the problem.

Another possibility if the VCR turns on but won't play tapes is that the tape-in switch is either broken or dirty. This switch is triggered when the tape is pushed into the VCR slot. After the switch is hit, the tape will pull down into playing position. If your VCR seems to be having this problem, take off the cover and place a tape in the slot with the power on. As with any piece of electronic equipment, be extremely careful when the cover is off and the power is on—you could take a nasty jolt (and trash the VCR, in this case) if you aren't. After you place the VHS tape in the slot, you should be able to find where the switch is. After you find it, you can take the tape out and try pushing the switch with your hand to see if it works. If not, you should be able to determine whether the problem is mechanical (the lever of the switch isn't turning the switch on) or electrical (the switch is broken).

Other common problems

Some other common problems with VCRs include nonfunctional or dirty knobs, buttons, and switches, and blown filter capacitors. These troubles are common to nearly all types of consumer electronics equipment. For more information on these topics, see the general repairs section near the end of this chapter.

TVs

Many, many different problems can occur in a TV, but troubleshooting is much more difficult now that very few discrete components are used. Instead of a capacitor or transistor blowing in a TV, chances are better that an IC failed. Now, rather than fix the TV at the component level, the damaged board is popped out and a new assembly is snapped in.

TV troubleshooting is different from troubleshooting any other type of electronics equipment. With most equipment, something fails and you have very few symptoms to determine what went wrong. In the case of the TV, just about every general form of circuit failure appears with a different effect on the video of the screen. Even if you can't troubleshoot or repair a TV with electronics equipment, you might be able to determine what circuits have failed just from looking at the picture. With this information, you should be able to help the TV repairer to save a little time with diagnosing the problem. Hopefully, presenting this little bit of information will also prevent unscrupulous repair shops from overcharging you.

Because of the high voltages required in televisions, the power circuitry is particularly vulnerable to damage. If the TV consistently blows fuses in your house, beware of failures in the power supply. In the past, an electrically hot TV also indicated power supply problems, but these days TVs are manufactured with insulating plastic cases and rubber or plastic buttons, so this symptom has virtually disappeared. Another symptom of a power supply problem is when the TV turns itself off immediately or after a few minutes of operation.

Another difficult fix is TV tubes. If you suspect that the CRT has failed in your TV, don't try to replace it yourself unless you are an electronics technician. First of all, replacement CRTs can be very difficult to find, particularly for TVs that are more than two years old. Also, new replacement CRTs are very expensive--having a CRT replaced at a repair shop will often cost more than the TV is worth. Finally, CRTs are very dangerous to replace. A vacuum exists in all electron tubes (CRTs included), so the tube will implode (rather than explode) if the glass is broken. We're not talking about an implosion of house-razing force here, but anytime that shattered glass is involved, the situation can be very dangerous (Fig. 12-11). And if you approach the CRT from the back side, be careful of the CRT neck, which is narrow and fragile.

The following list of symptoms is very basic and will help you to get started with the process. For actual repairing information with examples, see *Troubleshooting & Repairing Solid-State TVs (3rd Edition)* by Homer Davidson.

* *Horizontal lines* Power supply or horizontal oscillator.
* *Poor focus* Power supply or focus control.
* *Picture too wide* Power supply.
* *Intermittent picture* Loose connection somewhere.
* *Horizontal white line* Yoke-area problem.
* *Vertical white line* Yoke-area problem
* *Picture folds over horizontally* Pincushion or other problems.
* *Picture folds over vertically* Vertical output circuits.
* *Squished picture* Vertical output circuits.

Fig. 12-11 The glass neck of a CRT, showing the yoke-coil assembly (center of the picture). The neck is very fragile.

* *TV stations tough to tune in* Dirty mechanical tuner.
* *No GUI display* Tuner control module.
* *Colors missing* Color-output circuits or picture-tube circuits.
* *"Hum bars" appear* Power-supply circuits.
* *Colors not focusing* Color circuits need professional adjustment.
* *"Negative" picture* Picture tube circuits.
* *No tint control* Tint-control circuits.
* *Multiple images or "ghosting"* TV reception; not an equipment fault.
* *Snowy picture* Cable or TV antenna problem or IF circuit problem
* *No picture, no sound* Tuner or IF circuits.
* *Weak picture* Video circuits.
* *Brightness control problems (no control)* Brightness-control circuits.

* *Intermittent changes in brightness* Brightness-control circuits or CRT.
* *Smearing pictures* Video circuits.
* *Dark, unstable picture* AGC circuits.
* *Bad audio* Audio circuits.

AMPLIFIERS

Dedicated amplifiers are a difficult fix for the nontechnician. There are no moving parts and no belts or motors to maintain. Everything is entirely electrical, not mechanical. Because of the lack of moving parts, amplifiers rarely fail (unless they have been driven too hard) and as a result, the better manufacturers have warranties of several years on each amplifier. Normally, when something goes wrong, the amplifier fails entirely. Then, chances are you're stuck. You must take the amp to a factory service center because if you open up the amplifier to see what failed, the warranty will be voided.

If your amplifier no longer operates and the LED indicators don't even light, first check the plug or surge protector to be sure that nothing has become unplugged or that the surge protector has been turned off. In a rare instance, the line might have taken a nasty surge and has either shut off or totally blown the surge protector. See if the protector button will reset. If the surge protector is blown and won't reset, try plugging the amplifier back into the wall and see if it works.

If the On indicator is lit, but there is no audio, check the audio lines, connections, and amplifier front-panel buttons. It's common, especially in systems where you do lots of recording or switching components and audio cables, to have the cables switched around or not plugged in. My systems, for example, confuse most people: I don't have a multi-input A/V receiver, so I either use a mixer to switch between audio sources or I just switch cables. Then, I set my cassette deck on record and feed any audio into it, then on to the amplifier. Most people who try to listen to something on my stereo don't realize that the cassette deck is controlling the audio for the whole system and they can't get it running. This type of problem probably won't occur in your system, but if you have a family member who toys around with audio, look out!

If your amplifier appears to be dead or not working properly and you plan to take it in for repairs soon, wait a few days first. Check everything over a few times and give it a few days so that you can think of any simple means that you could use to correct the problem.

Unlike videocassette players, cassette decks, and turntables, which have moving parts and are subject to dirt and wear, dedicated amplifiers generally will not break down unless they have been abused. However, the next few sections cover a few problems that can occur.

GENERAL REPAIRS

The following repairs apply to a wide variety of audio equipment. For example, any piece of equipment that has a potentiometer knob could become scratchy. The last two problems, leaky capacitors and burned resistors, occur most often in tube amplifiers and old radios. These electronic repairs are all very simple and are intended for general hobbyists or audio fiends, not for those with an electronics background. As a result, the repairs require no equipment, aside from a soldering iron, some solder, and your senses. If you are really interested in repairing audio equipment, get some test equipment, such as a good DMM (digital multimeter), with a capacitance tester, transistor checker, etc., and check out some more in-depth troubleshooting and repair books.

"SCRATCHY" KNOBS

Sometimes knobs will become intermittent and make scratching noises as you turn them. For example, sometimes no sound is audible as you turn up the volume control. Then, suddenly, you might hear a loud scratching sound and the audio starts blaring through.

The problem is that the potentiometer (a variable resistor) contains metal contacts. These contacts slide against each other as you turn the knob. If the contacts corrode, tarnish, or become dirty, the potentiometer will become scratchy sounding and/or intermittent. The cure is simple. Spray some cleaning fluid inside of the potentiometer and turn the knob back and forth. If the spray is thoroughly worked over the contacts, they should become clean.

Nonfunctional buttons

Buttons sometimes fail after years of use. One common ailment is that a pushbutton will no longer stay set. These buttons are typically power-on/off buttons, which receive quite a bit of abuse over the years (although other pushbuttons, such as Noise Reduction buttons, can also be similarly damaged). If these buttons fail, you have two options: Go into the innards of the equipment and bypass the switch (effectively "hardwiring" the equipment "on," in the case of an on/off button), or try to find a replacement pushbutton that will fit the cabinet.

The problem with the first option is that you no longer have control over the equipment button. That's no big deal, if the pushbutton is an on/off switch; you can turn it on or off via a surge-suppressor switch. The problem with the second solution is that you probably won't find an exact replacement pushbutton for your application. With luck, you'll find one that fits the hole, but you probably won't find one that matches. This is no problem if your listening area is in a grungy room in the basement, but you will probably want to be more careful with equipment in a living room audio system.

If your particular piece of equipment isn't too old, you might be able to order replacement parts from the manufacturer's parts depot. Aside from this option, if you want to fix your equipment so that it looks "presentable" for your living room, you can search for an identical (but nonfunctional) piece of equipment at a hamfest or in some of the audio newsgroups on the Internet.

Leaky capacitors

Aside from the constantly degrading electron tubes that were used in 1930s through 1950s electronic equipment, most are also loaded with paper tube capacitors. These capacitors consist of a paper or bakelite tube that is filled with rolled sheets of aluminum foil and wax paper. In general, you can dig up just about any unrestored radio or amplifier from 40 years (or longer) ago, and at least one of the paper capacitors will be problematic (possibly causing the equipment not to work at all, or as is often the case, put a severe hum on the audio). Because of these problems, most equipment restorers suggest that you replace every paper capacitor with mylar capacitors. If you use the equipment regularly, it will save you lots of time to replace them all at once.

The next dilemma is equipment restoration. Many antique radio buffs do not believe that modern mylar capacitors should be seen within the case of a piece of antique audio equipment. Rather than make a wholesale replacement of these parts, they will place the much smaller mylar capacitors within the empty shells of paper capacitors. The techniques vary, but some people will place the old capacitors in boiling water to remove the old wax, then will push out the "guts" of the old paper capacitor. If you plan to use the old capacitor shells, be sure to check the types of capacitors and work with rubber gloves. Over the years, capacitors have used lots of different chemicals inside—many contained acid solutions or PCB-laden oil. Although paper capacitors aren't filled with oil, it's best to check with some knowledgeable antique electronics restorers before tearing into some of these components.

Aside from the paper capacitors, you also need to look over the old electrolytic capacitors. These capacitors are large metal or paper cylinders that contain a chemical solution (often nitric or sulfuric acid). These capacitors will often either dry up inside and fail or they will begin to leak. Either way, they render the equipment useless. Look for spots under the electrolytic capacitors on the chassis; they will either look wet or leave a lightly powdery white residue. If you find capacitors like these, immediately replace them, of course, with components that have the same values. And remember that electrolytic capacitors have polarity (a positive and a negative lead), so connect any replacements following the original polarity.

BURNED RESISTORS

A sure way to find failed components in a piece of equipment is to find parts that have burned. If it's burning, there's a problem, right? The two most common burn failures in a piece of tube electronic equipment are resistors and transformers. Transistors and ICs will fail in newer equipment, but you don't typically see the smoke or flames associated with resistor and transformer failures. Transformers fail with smoke or arcing sounds, but that's about it. Resistors are much more dramatic, so that's what is covered here.

Most resistors in electronic equipment are inexpensive carbon-composition units. If they receive too much voltage or suddenly fail, they will often smoke. I've even seen some resistors that had burned in half! To find these failures, look the chassis or PC board over for burns or ashes, and look for burned resistors. If a resistor has burned up, the unit will most likely not work at all. Be sure to replace it with a resistor that has a similar resistance and power rating. Then test it. If it works, leave it on for a while to see if the resistor blows again. Burned up resistors are often signs of other component failures.

CONCLUSION

Although the electronics layperson is severely limited by what he or she can do to repair the equipment in a home theater system, it is possible to perform some simple maintenance and repair procedures that can save quite a bit of money and prolong the life of your system.

A

Sources

ABC
130 Hwy. 33, Englishtown, NJ 07726 USA
(800) 354-1324 (tel)
(908) 780-6600 (tel)
(908) 294-7480 (tel)
Sells home and amateur audio and video equipment, such as camcorders, VCRs, speakers, cassette decks, TVs, home theater equipment, etc.

Abe's of Maine
1957 Coney Island Ave., Brooklyn, NY 11223 USA
(800) 545-2237 (tel)
(800) 992-2379 (service tel)
(718) 645-1818 (tel)
(718) 998-5216 (order fax)
(718) 382-8596 (service fax)
Sells professional and home photographic audio and video equipment, such as cameras, camcorders, CD players, cassette decks, video editors, monitors, VCRs, cables, etc. Offers a free catalog.

Audio Visionaries
3650 Austin Bluffs, Colorado Springs, CO USA
(719) 532-1515 (tel)
(888) 532-1518 (tel)
http://www.audiovisionaries.com
Offers a huge variety of home theater equipment—everything from VCRs, TVs, and speakers to theater seats.

Accurate Audio Video
2301 N. Central, Ste. 182, Plano, TX 75075 USA
(800) 414-1849 (tel)
Sells a wide variety of home audio and video equipment.

AITech
47971 Fremont Blvd., Fremont, CA 94538 USA
(800) 882-8184 (tel)
(510) 226-9169 (tel)
(510) 226-8996 (fax)
Manufactures scan converter boards to display computer video on televisions.

Alpha Stereo
57 Smithfield Blvd., Plattsburgh, NY 12901 USA
(518) 561-2822 (tel)
(518) 561-2961 (fax)
Visa, MasterCard, Discover, American Express
Sells audio and video equipment, such as camcorders, VCRs, speakers, cassette decks, TVs, home theater equipment, etc.

Audioquest
915 Calle Amanecer #O, San Clemente, CA 92673 USA
(714) 498-2770 (tel)
(714) 498-5112 (fax)
Manufactures audio and power cables for high-end, amateur, and professional audio and video applications.

audio studio
303 Newbury St., Boston, MA 02115 USA
(617) 267-1001 (tel)
414 Harvard St., Brookline, MA 02146 USA
(617) 277-0111 (tel)
(617) 277-2415 (fax)
Sells home and amateur audio and video equipment, such as camcorders, VCRs, speakers, cassette decks, TVs, home theater equipment, etc.

AVerMedia
47923A Warm Springs

Blvd., Fremont, CA 94539 USA
(510) 770-9899 (tel)
(510) 770-9901 (fax)
(604) 278-3224 (Canada tel)
(604) 278-2909 (Canada fax)
Manufactures PC/Mac-to-TV converters for desktop video applications.

Beach
203 Rt. 22 E., Greenbrook, NJ 08812 USA
(800) 634-1811 (tel)
(908) 424-1101 (info tel)
(908) 424-1105 (fax)
(908) 424-1103 (service tel)
Sells high-end home, amateur, and professional audio and video equipment, such as camcorders, video editors, karaoke players, VCRs, cables, etc.

Berger Brothers
209 Broadway (Rt. 110), Amityville, NY 11701 USA
(800) 262-4160 (tel)
(516) 264-4160 (tel)
(516) 264-1007 (fax)
Sells a wide variety of photographic and video equipment, including cameras, camcorders, VCRs, etc. Offers a full money-back guarantee and a 240-page catalog.

Bri-Tech A/V
3900 Veterans Hwy., Ste. 120, Bohemia, NY 11716 USA
(800) 467-7707 (tel)
Visa, MasterCard, Discover, American Express
Sells a wide variety of new and used high-end home audio and video equipment.

C & A Audio
193 Bellevue Ave., Upper Montclair, NJ 07043 USA
(201) 744-0600 (tel)
Sells a wide variety of home audio and video equipment.

Camera Sound
Rt. 252 S., Berwyn, PA 19312 USA
(800) 477-0022 (tel)
(215) 949-2124 (tel)
(215) 949-1085 (fax)
1104 Chestnut St., Philadelphia, PA 19312 USA
110 Lincoln Hwy., Fairless Hills, PA 19030 USA
(215) 949-1086 (tel)
Sells home, amateur, and professional audio and video equipment, such as camcorders, video editors, cassette decks, VCRs, receivers, etc.

Camera World of Oregon
500 S.W. 5th Ave., Portland, OR 97204 USA
(800) 729-8937 (tel)
(800) 729-8940 (free catalog tel)
(503) 227-6008 (tel)
(503) 222-7070 (fax)
Sells a wide variety of photographic, audio, video, and office equipment, including cameras, lenses, camcorders, video editors, telephones, TVs, receivers, karaoke decks, ministereos, etc. Offers a free catalog.

Citizen
2450 Broadway, #600, Santa Monica, CA 90411 USA
(800) 421-6516 (tel)
(310) 453-0614 (tel)
(310) 453-2814 (fax)
Manufactures color LCD camcorder monitors for professional, amateur, and home video recording applications.

Coast to Coast
2570 86th St., Brooklyn, NY 11214 USA
(800) 788-5555 (tel)
(718) 265-1723 (service tel)
(718) 265-6800 (info tel)
(718) 265-1754 (fax)
Sells home and amateur audio and video equipment, such as camcorders, video editors, monitors, VCRs, cassette decks, car audio equipment, etc. Offers extended warranties.

Colonel Video & Audio
16451 Space Center Blvd., Houston, TX 77058 USA
(800) 423-VCRS (orders tel)
(800) Video-97 (repeat callers tel)
(713) 486-8866 (cust serv tel)
(713) 486-8300 (fax)
Sells home, amateur, and professional audio and video equipment, such as camcorders, VCRs, video editors, monitors, TVs, home theater equipment, etc.

ComputAbility Consumer Electronics
P.O. Box 17882, Milwaukee, WI 53217 USA
(800) 554-9954 (tel)
(414) 357-8181 (tech tel)
(414) 357-7814 (fax)
http://www.compuability.com
Sells home, amateur, and professional audio video equipment, such as camcorders, VCRs, video editors, microphones, TVs,

home theater equipment, etc.

Computer Friends
14250 NW Science Park Dr., Portland, OR 97229 USA
(800) 547-3303 (tel)
(503) 626-2291 (tel)
(503) 643-5379 (fax)
cfriends@teleport.com
Sells computer scanners, printers, VGA-to-TV encoders, etc., for professional and personal desktop artwork, publishing, and video applications.

Comtrad Industries
2820 Waterford Lake Dr., #106, Midlothian, VA 23113 USA
(800) 992-2966 (tel)
Sells wireless video and audio transmitter/receivers to couple VCRs with TVs in other locations, etc.

Connectix
2655 Campus Dr., San Mateo, CA 94403 USA
(800) 839-3628 (tel)
(415) 571-5100 (tel)
(415) 571-5195 (fax)
info@connectix.com
Manufactures tiny video cameras for a wide variety of computer applications, including video-conferencing, movie making, etc.

Connectors Unlimited
P.O. Box 5983, Manchester, NH 03108 USA
(800) 549-5955 (tel)
(603) 641-1179 (fax)
Sells a wide variety of connectors for audio, video, cellular, CATV, data, transmitting, etc.

Cool-Lux
409 Colle San Pablo, Ste. 105, Camarillo, CA 93012 USA
(800) 223-2589 (tel)
(805) 482-4820 (tel)
(805) 482-0736 (fax)
Manufactures and sells a variety of lights for professional and amateur video recording and batteries and battery belts, especially for camcorders. Offers a free catalog.

Damark
7101 Winnetka Ave. N., P.O. Box 29900, Minneapolis, MN 55429 USA
(800) 827-6767 (tel)
Sells a wide variety of new and refurbished electronics, including car and home stereo equipment, home and semiprofessional video equipment, personal computers, office equipment, etc.

Data Translation
100 Locke Dr., Marlboro, MA 01752 USA
(800) 525-8528 (tel)
(508) 481-8620 (fax)
574-7126@mcimail.com
Manufactures video capture boards and video production systems for personal and professional computer video applications.

DataVision
445 Fifth Ave., New York, NY 10016 USA
(800) 482-7466 (tel)
(212) 825-1823 (customer service tel)
(212) 689-1111 (tel)
(212) 825-1970 (fax)
datavis@aol.com
Sells home, amateur, and professional audio and video equipment and computers, such as camcorders, video editors, battery packs, VCRs, cables, etc.

DAVES (Direct Audio Video Entertainment Systems)
441 Kearny Ave., Kearny, NJ 07032 USA
(800) 308-0354 (tel)
(201) 997-7812 (tel)
Sells home and amateur audio and video equipment, such as camcorders, VCRs, speakers, cassette decks, TVs, home theater equipment, etc.

Definition Audio Video
2901 W. 182nd St., Redondo Beach, CA 90278 USA
(310) 371-0019 (tel)
Sells a wide variety of home theater equipment.

Digicomm International
14 Inverness Dr. E., #A124, Englewood, CO 80112 USA
(303) 799-3444 (tel)
(303) 799-9366 (fax)
Manufactures video titlers for professional TV broadcast and video production applications.

Digital Processing Systems
(606) 371-5533 (tel)
Manufactures and sells PC computer-based video recorders for home and professional video production applications.

Dobbs-Stanford
2715 Electronic Ln., Dallas,

TX 75220 USA
(214) 350-4222 (tel)
Manufactures computer-to-video converters for professional, amateur, or home video playback applications.

Dolgin Engineering
(617) 926-2608 (tel)
Manufactures S-VHS junction boxes for home, amateur, and professional video editing applications.

D-Vision Systems
8755 W. Higgins Rd., 2nd Fl., Chicago, IL 60631 USA
(800) 838-4746 (tel)
(312) 714-1400 (tel)
Manufactures computerized digital editing systems for home and professional video applications.

Eagle Distributors
2438 Albany St., Kenner, LA 70062 USA
(800) 726-5930 (tel)
(504) 464-5930 (tel)
(504) 468-9741 (fax)
Sells replacement parts and accessories for audio and video equipment.

Electronic Mailbox
10-12 Charles St., Glen Cove, NY 11542 USA
(800) 323-2325 (tel)
(516) 759-1943 (tech advice tel)
(516) 671-3092 (fax)
Sells amateur and professional video editing and production equipment, including camcorders, batteries, wireless microphones, scan converters, video titlers, A/V processors, desktop video boards, etc. Offers a 10-day warranty.

Elite Video
321 Ouachita Ave., Hot Springs, AR 71901 USA
(800) 468-1996 (tel)
(800) 662-8125 (tel)
(501) 321-0440 (tel)
(501) 321-2327 (fax)
jonvideo@aol.com
Manufactures video mixers, broadcast video processors, and video duplicators for home video and professional/broadcast video applications. Offers a free catalog.

Emerald Vision & Sound
127 W. 24th St., New York, NY 10011 USA
(800) 980-2929 (tel)
(212) 463-7663 (tel)
(212) 463-7143 (fax)
Sells a wide variety of video equipment, including camcorders, video editors, VCRs, battery chargers, microphones, etc.

Executive Photo and Electronics
120 W. 31st St., New York, NY 10001 USA
(800) 223-7323 (order tel)
(800) 882-2802 (service tel)
(212) 947-5290 (order tel)
(212) 947-5295 (service tel)
(212) 239-7157 (fax)
Sells home, amateur, and professional photographic, audio, video, and office equipment, such as camcorders, VCRs, car audio equipment, shortwave radios, fax machines, etc.

Express Computer & Video
726 Seventh Ave., New York, NY 10019 USA
(800) 879-5244 (tel)
(212) 397-6700 (tel)
(212) 397-6848 (fax)
Sells home and amateur audio and video equipment, such as camcorders, VCRs, TVs, TV projectors, home theater equipment, etc.

Factory Direct
35 W. 35th St., New York, NY 10001 USA
(800) 695-3535 (tel)
(212) 564-4399 (tel)
(212) 564-4398 (customer service tel)
(212) 643-9727 (fax)
Sells home, amateur, and professional audio and video equipment, such as camcorders, VCRs, video editors, monitors, TVs, home theater equipment, etc.

Family Photo & Video
1961 Coney Island Ave., Brooklyn, NY 11223 USA
(800) 899-7468 (tel)
(800) 405-7468 (customer service tel)
(718) 645-1298 (tel)
(718) 998-6650 (fax)
Sells home, amateur, and professional audio and video equipment, such as camcorders, VCRs, video editors, monitors, karaoke decks, home theater equipment, etc. Offers a free catalog.

Fast Electronic US
1 Twin Dolphin Dr., Redwood City, CA 94065 USA
(800) 684-MOVIE (tel)
(415) 802-0746 (fax)
Manufactures and sells digital video editing packages for home, amateur, and professional video editing on computer.

Free Trade Video
4718 18th Ave., #127,
Brooklyn, NY 11201 USA
(800) 234-8813 (tel)
(800) 455-6677 (tel)
(718) 435-4151 (tel)
(718) 633-6890 (fax)
Sells home, amateur, and professional audio and video equipment, such as camcorders, video editors, microphones, VCRs, cables, etc. Offers a 15-day return policy.

Future Video Products
28 Argonaut, #140, Aliso Viejo, CA 92656 USA
(714) 770-4416 (tel)
Manufactures edit controllers for home, amateur, and professional video editing applications via computer.

Gabriel Video & Camera
260 W. Swedesford Rd., Berwyn, PA 19312 USA
(800) 454-1010 (tel)
(610) 889-3960 (tel)
(610) 647-5830 (fax)
Sells home, amateur, and professional audio and video equipment, such as camcorders, video editors, microphones, VCRs, cables, etc.

GE
5365 A-2 Robin Hood, Norfolk, VA 23513 USA
(804) 244-3226 (tel)
Appliances/Microwave Products
Bldg. 41, Rm. 106, Louisville, KY 40225 USA
(502) 452-3568 (tel)
Manufactures a wide variety of electronics equipment, including home audio equipment (cassette players, etc.), home video equipment (VCRs, etc.), office equipment (telephones, answering machines, etc.), and more.

Genesis Camera
814 W. Lancaster Ave., Bryn Mawr, PA 19010 USA
(800) 575-9977 (tel)
(610) 527-5261 (tel)
(610) 527-5148 (fax)
Sells home, amateur, and professional video equipment, such as camcorders, VCRs, video editors, microphones, video switchers, mixers, etc.

Genoa Systems
(408) 362-2900 (tel)
(408) 432-8342 (tech support tel)
(408) 434-0997 (fax)
(408) 943-1256 (BBS)
Manufactures graphics boards, audio boards, tape drives, video capture boards, etc., for personal computer applications.

G&G Technologies
350 N. St., Teteboro, NJ 07608 USA
(800) VIDEO-911 (tel)
(201) 288-8900 (tel)
(201) 288-6881 (fax)
Sells a variety of home, amateur, and professional video equipment, including VCRs, A/V mixers, camcorders, and video editors.

Glidecam Industries
676 State Rd., Plymouth, MA 02360 USA
(800) 949-2089 (tel)
http://www.glidecam.com
Manufactures Glidecam videocamera stabilizers for home, amateur, and professional video recording applications.

Global Video
260 W. Swedeford Rd., Berwyn, PA 19312 USA
(800) 420-5050 (tel)
(610) 889-3960 (tel)
(610) 647-5830 (fax)
Sells home audio, home video, and professional video equipment, such as camcorders, video editors, CD players, VCRs, cables, etc.

Goldstar Electronics International
1000 Sylvan Ave., Englewood Cliffs, NJ 07632 USA
(201) 816-2000 (tel)
(201) 816-0636 (fax)
Parts Depot
201 James Record Rd., Huntsville, AL 35824 USA
(800) 222-6457 (consumer tel)
(800) 562-0244 (dealers & tech support tel)
(800) 448-4026 (fax)
(205) 772-8860 (tel)
(205) 772-8987 (fax)
Manufactures camcorders and VCRs for home, amateur, and professional video editing, recording, and viewing applications.

Go Video
14455 N. Hayden Rd., #219, Scottsdale, AZ 85260 USA
(602) 998-3400 (tel)
(602) 998-8312 (fax)
Manufactures dual-deck VCRs for home and professional video applications.

Home Theatre Systems
44 Rt. 23N, Little Falls, NJ 07424 USA
(800) 978-7768 (tel)
(201) 890-5959 (tech support tel)
(201) 890-9142 (fax)
Sells a wide variety of home audio and home theater equipment, including audio and video speakers, receivers, video projectors, TVs, VCRs, laser-discs, CD players, preamps, power amplifiers, cassette decks, satellite dishes, and accessories

J & R Music World
59-50 Queens Midtown Expwy., Maspeth, Queens, NY 11378 USA
(800) 221-8180 (tel)
Sells most everything electronic, including audio and video equipment, computers, office equipment, and much more. Offers a free catalog.

JVC
41 Slater Dr., Elmwood, NJ 07407 USA
(800) 526-5308 (tel)
(201) 794-3900 (tel)
(201) 797-0919 (fax)
Manufactures a wide variety of home audio and video equipment, including VCRs, A/V receivers, camcorders, etc.

Kenwood
P.O. Box 22745, 2201 E. Dominguez St., Long Beach, CA 90801 USA
(800) 950-5005 (tel)
(213) 639-9000 (tel)
(310) 639-4200 (tel)
(310) 761-8246 (fax)
(310) 609-2127 (fax)
6070 Kestrel Rd., Mississauga, ON L5T 1S8 Canada
(310) 639-5300 (customer support/brochure tel)
(800) KENWOOD (service locations/parts tel)
(310) 761-8284 (BBS)
Manufactures amateur radio equipment (including transceivers, receivers, amplifiers, two-ways, and accessories) and stereo equipment (including receivers, speakers, cassette decks, CD changers, car head units, etc.).

Magnavox
1 Philips Dr., Knoxville, TN 37914 USA
(800) 835-3506 (tel)
(615) 521-4391 (tel)
(432) 475-0317 (tel)
http://www.magnavox.com
Manufactures monitors for personal desktop computer systems; CD players, TVs, cassette decks, loudspeakers, VCRs, etc., for home video and home audio applications; etc.

Marine Park Camera & Video
3126 Ave. U., Brooklyn, NY 11229 USA
(800) 448-8811 (tel)
(800) 360-1722 (order inquiry)
(718) 891-1878 (tel)
Sells home, amateur, and professional audio and video equipment, such as camcorders, VCRs, video editors, speakers, TVs, home theater equipment, etc.

Markertek Video Supply
4 High St., Saugerties, NY 12477 USA
(800) 522-2025 (tel)
(914) 246-3036 (tel)
(914) 246-1757 (fax)
Sells a wide variety of equipment (including standards converters, camcorders, video editors, digital standards converters, etc.) for professional audio and video applications. Offers a free 174-page catalog.

Matsushita (Panasonic)
50 Meadowland Pkwy., Seacaucus, NJ 07094 USA
(201) 348-7000 (tel)
(201) 348-7527 (fax)
Manufactures head units, loudspeakers, power amplifiers, etc,. for car stereo applications; camcorders, televisions, VCRs, etc,. for home video applications; cassette decks, CD players, amplifiers, etc., for home audio applications, etc.

Midwest Micro
6910 U.S. Rt. 36 E., Fletcher, OH 45326 USA
(800) 235-8940 (tel)
(800) 262-6622 (tech supp tel)
(800) 728-8584 (cust serv tel)
(800) 883-1461 (dealers tel)
(800) 562-6622 (fax)
(800) 445-2025 (info fax)
Manufactures and sells personal desktop and laptop computers, printers, and scanners; also sells computer parts and boards, such as floppy drives, hard drives, monitors, processor upgrades, video cards, modems, keyboards, etc. Offers a 30-day money-

back guarantee and a free catalog

Minolta
101 Williams Dr., Ramsey NJ 07446 USA
(800) 9-MINOLTA (tel)
(201) 825-4000 (tel)
Manufactures cameras, fax machines, digital systems, imagers, camcorders, photocopiers, etc., for a wide variety of photographic, office, video applications.

Mitsubishi Electronics America
6100 Atlantic Blvd., Norcross, GA 30071 USA
5757 Plaza Dr., P.O. Box 6007, Cypress, CA 90630 USA
(800) 828-6372 (tel)
(800) 553-7278 (tel)
(800) 843-2515 (Display Products tel)
(800) 937-0000 X435 (tel)
(800) 387-9630 (Canada Display Products tel)
(714) 220-2500 (tel)
(800) 937-2094 (tel)
(800) 825-6655 (fax)
(714) 220-4792 (fax)
http://www.directnet.com/mitsubishi
Manufactures VCRs for home, amateur, and professional video editing, recording, and viewing applications; manufactures integrated circuits for a wide variety of electronics applications.

Modern Electronics
2125 S. 156th Circle, Omaha, NE 68130 USA
(800) 906-6664 (tel)
Sells cable TV converters, descramblers, video stabilizers, filters, etc. Offers a money-back guarantee, a 30-day trial, and a free catalog.

MusiCraft
1280 E. Dundee Rd., Palantine, IL 60067 USA
(708) 991-7220 (tel)
1800 S. Halsted, Homewood, IL 60430 USA
(708) 799-2400 (tel)
Sells home and amateur audio and video equipment, such as camcorders, VCRs, speakers, cassette decks, TVs, home theater equipment, etc.

National Recording Supplies
764 5th Ave., Brooklyn, NY 11232 USA
(800) 538-2336 (tel)
(718) 369-8273 (tel)
(718) 369-8275 (fax)
Sells a variety of recording supplies, including bulk audio and video cassettes, multitrack tape, etc., for professional recording studios. Offers a free catalog.

New West Electronics
4120 Meridian, Bellingham, WA 98226 USA
(800) 488-8877 (tel)
(360) 734-3314 (fax)
Sells home, amateur, and professional audio and video equipment, such as camcorders, VCRs, video editors, monitors, TVs, home theater equipment, etc.

North Country Radio
P.O. Box 53, Wykagyl Station, New Rochelle, NY 10804 USA
(914) 235-6611 (tel)
(914) 576-6051 (fax)
(518) 854-9280 (technical info)
Manufactures and sells radio/audio/video kits, including amateur TV transmitters and receivers, etc.

Now! Recording Systems
32 W. 39th St., 9th Fl., New York, NY 10018 USA
(800) 859-2579 (tel)
(212) 768-7800 (tel)
(212) 768-9740 (fax)
Sells blank audio and video recording cassettes, etc. Offers a free catalog; minimum $25 order.

One Call West
418 W. Riverside, Spokane, WA 99201 USA
(800) 633-5255 (tel)
(509) 838-4387 (fax)
onecall@iea.com
Sells home and amateur audio and video equipment, such as camcorders, VCRs, speakers, car audio equipment, TVs, home theater equipment, etc.

Paramount Video & Audio
110 Lincoln Hwy., Box 15, Langhorne, PA 19030 USA
(800) 477-0330 (tel)
(215) 949-2860 (tel)
(215) 949-1085 (fax)
Sells high-end home, amateur, and professional audio and video equipment, such as camcorders, VCRs, video editors, monitors, TVs, CD and cassette decks, etc.

PC Video Conversion
1340 Tully Rd., #309, San

Jose, CA 95122 USA
(408) 279-2442 (tel)
(408) 279-6105 (fax)
Manufactures computer converter boards to produce NTSC, PAL, S-VHS, RGB, etc., video from a computer.

The PC Zone
15815 SE 37th St., Bellevue, WA 98006 USA
(800) 258-2088 (tel)
(206) 603-2500 (fax)
Sells personal desktop and laptop computers, printers, software, and scanners; also sells computer parts and boards, such as floppy drives, hard drives, monitors, processor upgrades, video cards, modems, keyboards, etc.

Peach State Photo
1706 Chantilly Dr. NE, Atlanta, GA 30324 USA
(800) 766-9653 (tel)
(404) 633-2699 (cust serv tel)
(404) 321-6316 (fax)
Sells home, amateur, and professional photographic, audio, and video equipment, such as cameras, lenses, camcorders, VCRs, video editors, monitors, tripods, etc.

Philips Consumer Electronics
1 Philips Dr., Knoxville, TN 37914 USA
(800) 822-4788 (tel)
(800) 235-7373 (tel)
(615) 521-4316 (tel)
(615) 521-4391 (tel)
(800) 522-7464 (tel)
(315) 682-9006 (fax)
Service Department
401 Old Andrew Johnson Hwy., Jefferson City, TN 37760 USA
(800) 851-8885 (replacement parts tel)
(615) 475-8869 (tech tel)
(800) 535-3715 (fax)
Broadband Networks division
(800) 448-5171 (tel)
(800) 522-7464 (tel)
(315) 682-9006 (fax)
http://pps.philips.com
Manufactures and sells a variety of consumer audio and video electronics, including TVs, VCRs, home stereo equipment (including amplifiers, cassette decks, CD players, DCC decks, etc.), car stereo equipment (including head units, amplifiers, speakers, etc.), cable-TV equipment, computer equipment (including CD-ROM recorders), and shortwave radios.

Philly's Camera
27 S. 11th St., Philadelphia, PA 19107 USA
(800) 923-2022 (tel)
(215) 922-5130 (tel)
(215) 922-5135 (fax)
Sells a variety of optical, audio, and video equipment, including cameras, camcorders, CD players, receivers, cassette decks, video editors, VCRs, etc.

Planet Electronics
8418 Lilley Rd., Canton, MI 48187 USA
(800) 247-4663 (tel)
(313) 453-4750 (tel)
(313) 453-4750 (fax)
Visa, MasterCard, Discover
Sells home, amateur, and professional audio and video equipment, such as camcorders, VCRs, video editors, monitors, TVs, home theater equipment, etc. Offers a free 64-page catalog.

Powervideo
6808 Hornwood Dr., Houston, TX 77074 USA
(800) 877-2900 (tel)
(713) 772-4400 (tel)
(713) 772-6500 (fax)
Sells home, amateur, and professional audio and video equipment, such as camcorders, video editors, speakers, VCRs, cables, etc.

PrimeTime Video and Cameras
1104 Chestnut St., Philadelphia, PA 19107 USA
(800) 477-8445 (tel)
(215) 627-1080 (tel)
(215) 627-5886 (fax)
Sells home, amateur, and professional audio and video equipment, such as camcorders, laserdisc players, computer editing equipment, VCRs, accessories, etc.

Princeton Graphics
2801 S. Yale St., Ste. 110, Santa Ana, CA 92704 USA
(714) 751-8405 (tel)
(714) 751-5736 (fax)
http://www.prgr.com
Manufactures a variety of computer monitors.

Profeel Marketing
42 Main St., #109, Monsey, NY 10952 USA
(800) 776-3335 (tel)
(914) 425-2070 (tel)
(914) 425-0221 (fax)
Sells home, amateur, and professional audio and video equipment, such as

camcorders, video editors, monitors, VCRs, cables, etc.

PRV Sales
894 Green Pl., Woodmere, NY 11598 USA
(800) 670-6555 (tel)
Buys and sells new and lightly used video equipment.

Quasar
1707 N. Randall Rd., Elgin, IL 60123 USA
(708) 468-5600 (tel)
Manufactures televisions for home video applications.

RC Distributing
PO Box 552, South Bend, IN 46624 USA
(219) 236-5776 (tel)
Manufactures and sells video sync generators, satellite antennas, and satellite systems. Offers free information packets.

Recoton
46-23 Crane St., LIC, NY 11101 USA
(718) 392-6442 (tel)
2950 Lake Emma Rd., Lake Mary, FL 32746 USA
(800) 223-6009 (tel)
Manufactures antennas, blank audio tapes, connectors, microphones, loudspeakers, surge protectors, voltage converters, etc., for a wide variety of audio, video, office, and computer applications.

Reference Audio Video
18214 Dalton Ave., Gardena, CA 90248 USA
(310) 517-1700 (tel)
(310) 517-1732 (fax)
Sells a wide variety of home audio and video equipment.

Runco
26203 Production Ave., #8, Hayward, CA 94545 USA
(510) 293-9154 (tel)
(510) 293-0201 (fax)
http://www.runco.com
runco@runco.com
Manufactures laserdisc players, scan converters, digital video projectors, etc., for home theater applications.

Samman's Electronics
1166 Hamburg Tnpke., Wayne, NJ 07470 USA
(800) 937-3537 (tel)
(201) 696-6531 (tel)
Sells amateur and home audio and video equipment, such as camcorders, VCRs, video editors, receivers, monitors, TVs, home theater equipment, car audio equipment, etc.

Savant Audio
287 Clarksville Rd., Princeton Junction, NJ 08550 USA
(800) 628-0627 (tel)
(609) 799-9664 (tel)
(609) 799-8480 (fax)
pp000792@interramp.com
Sells home and amateur audio and video equipment, such as camcorders, VCRs, speakers, cassette decks, TVs, home theater equipment, etc.

Select Solutions
P.O. Box 6512, Champaign, IL 61826 USA
(800) 4-SNAPPY (tel)
(800) 322-1261 (tel)
(217) 355-2785 (customer service tel)
(217) 356-4312 (fax)
solutions@dvideo.com
Specializes in software and equipment for computer video applications, such as sound and video cards, hard drives, Bernoulli drives, etc.

6th Avenue Electronics
1030 6th Ave., New York, NY 10018 USA
(800) 394-6283 (tel)
(201) 467-0100 (tel)
331 Rt. 4W, Paramus, NJ 07652 USA
22 Rt. 22W, Springfield, NJ 07081 USA
Sells a wide variety of home video and audio equipment, including camcorders, VCRs, TVs, CD/laserdisc players, karaoke, home theater systems, etc.

Smile Photo & Video
29 W. 35th St., New York, NY 10001 USA
(800) 516-4202 (tel)
(212) 967-5900 (tel)
(800) 699-2836 (fax)
(212) 967-5912 (fax)
Sells home, amateur, and professional video equipment, such as camcorders, VCRs, video editors, monitors, TVs, PAL equipment, LCD projectors, etc.

Sony Electronics
1 Sony Dr., Park Ridge, NJ 07656 USA
(800) 352-7669 (tel)
(201) 930-1000 (tel)
National Parts Center
8281 NW 107th Terrace, Kansas City, MO 64153 USA
(816) 891-7550 (tel)
Service company
120 Interstate N. Pkwy. E.,

#414, Atlanta, GA 30339 USA
(404) 953-5862 (tel)
Service company
1200 N. Arlington Heights Rd., Itasca, IL 60143 USA
(708) 773-7500 (tel)
Service company
3201 Premier Dr., #100, Irving, TX 75063 USA
(214) 550-5270 (tel)
Service company
2729 Bristol St., Costa Mesa, CA 92626 USA
(714) 549-0294 (tel)
Publications department
P.O. Box 20407, Kansas City, MO 64195 USA
(816) 891-7550 (tel)
(800) 821-5662 (fax)
Manufactures digital signal processors for professional studio recording; integrated circuits for a wide variety of electronics applications; audio equipment (including tape decks, turntables, portable stereos, loudspeakers, car stereo equipment, etc); video equipment (including TVs, VCRs, video disc players, etc.); computer equipment (monitors, multimedia equipment, data storage devices, etc.); office equipment (including telephones, etc.); etc.

Sound City
Meadtown Shopping Center, Rt. 23 S., Kinnelon, NJ 07405 USA
(800) 542-7283 (order tel)
(800) 233-4010 (service tel)
(201) 838-9100 (tel)
Sells home, amateur, and professional audio and video equipment, such as camcorders, VCRs, video editors, monitors, TVs, home theater equipment, car audio equipment, etc. Offers a free 100-page catalog.

Sound Investment
3586 Pierce Dr., Chamblee, GA 30341 USA
(800) 659-8273 (tel)
(404) 458-1679 (tel)
(404) 458-0276 (fax)
Specializes in selling a variety of blank audio/video recording tapes for home, amateur, and professional applications.

The Stereo Trading Outlet
320 Old York Rd., Jenkintown, PA 19046 USA
(215) 886-1650 (tel)
Buys and sells a variety of used home audio and video equipment.

Supreme
1562 Coney Island Ave., Brooklyn, NY 11230 USA
(800) 332-2661 (tel)
(800) 241-2661 (Canada tel)
(718) 692-4140 (tel)
(718) 692-4760 (fax)
Sells home, amateur, and professional audio and video equipment, such as camcorders, VCRs, video editors, monitors, TVs, home theater equipment, etc.

Swagur Enterprises
P.O. Box 620035, Middleton, WI 53562 USA
(608) 592-7409 (tel/fax)
Sells a variety of satellite radio/video receiving equipment, including feed horns and dishes.

S&W Computers & Electronics
31 W. 21 St., New York, NY 10010 USA
(800) 874-1235 (tel)
(212) 463-8330 (tel)
(212) 463-8335 (fax)
Sells home audio and video, office, and computer equipment, such as camcorders, VCRs, cordless telephones, TVs, answering machines, etc.

Tape World
220 Spring St., Butler, PA 16003 USA
(800) 245-6000 (tel)
(412) 283-8298 (fax)
Sells more than 500 different types of blank recording cassettes, DATs, videotapes, MiniDiscs, etc. Offers a free catalog.

Ten Lab
2217 Malcolm Ave., Los Angeles, CA 90064 USA
(310) 474-6909 (tel)
(310) 441-1807 (fax)
http://www.tenlab.com
Manufactures digital video converters, rear projectors, etc., for home theater applications.

Tri State Camera & Video
650 6th Ave., New York, NY 10011 USA
(800) 537-4441 (tel)
(212) 633-2290 (tel)
(212) 633-7718 (fax)
Sells home, amateur, and professional audio, video, computer, and office equipment, such as camcorders, VCRs, video editors, monitors, TVs, fax machines, etc.

Uncle's Stereo
581 Broadway, New York, NY 10012 USA

(800) 978-6253 (tel)
(800) 924-2083 (tel)
(212) 343-9111 (tel)
(212) 343-9142 (fax)
Sells a wide variety of audio and video equipment, including DSPs, CD players, amplifiers, receivers, laser-disc players, loudspeakers, VCRs, TVs, etc.

Universal Video & Camera
1104 Chestnut St., Philadelphia, PA 19107 USA
(800) 477-1003 (tel)
(215) 889-7628 (tel)
(215) 647-5830 (fax)
Sells home, amateur, and professional audio and video equipment, such as camcorders, titlers, VCRs, video lights, etc.

USA Direct
894 Green Place, Woodbury, NY 11958 USA
(800) 830-1515 (tel)
Sells home, amateur, and professional audio and video equipment, such as camcorders, VCRs, CD players, TVs, home theater equipment, etc.

Video Discount Warehouse
295 Greewich St., #360, New York, NY 10007 USA
(800) 301-0028 (tel)
(212) 432-6058 (tel)
(212) 432-6104 (fax)
Sells home, amateur, and professional video equipment, such as camcorders, VCRs, video editors, etc. Offers a 7-day money-back guarantee.

Video Necessities
P.O. Box 411, Homecrest Stn., Brooklyn, NY 11229 USA
(800) 228-8480 (tel)
(718) 692-4512 (tel)
(718) 951-0843 (customer service tel)
(718) 951-9606 (fax)
Sells home, amateur, and professional audio and video equipment, such as camcorders, VCRs, video editors, monitors, TVs, home theater equipment, etc.

Videonics
1370 Dell Ave., Campbell, CA 95008 USA
(800) 338-EDIT (tel)
(408) 866-8300 (tel)
Manufactures editing suites for home video editing applications.

Vidikron
150 Bay St., Jersey City, NJ 07302
(201) 420-6666 (tel)
Manufactures projection TVs for home theater applications. Offers a free catalog.

Virtual Vision Sport
4132 E. Speedway Blvd., Tuscon, AZ 05712 USA
(800) 729-1020 (tel)
Manufactures a projection TV built into tinted goggles for home audio and video applications.

Visual Circuits
3989 Central Ave NE, Ste. 630, Minneapolis, MN 55421 USA
(612) 781-2186 (tel)
http://www.vcircuits.com
Manufactures and sells MPEG playback boards for amateur and professional video production applications.

World Trade Video
7 Broadway #1048, New York, NY 10004 USA
(800) 253-2639 (tel)
(212) 432-3156 (tel)
Sells home, amateur, and professional video equipment, such as camcorders, VCRs, video editors, monitors, etc.

XLO Electric
9480 Utica Ave., #612, Rancho Cucamonga, CA 91730 USA
(909) 466-0382 (tel)
(909) 466-3662 (fax)
Manufactures cables for audio/video applications.

INDEX

ABOUT THE AUTHOR

One of the "brightest stars" in hobby electronics writing, according to *Monitoring Times*, Andrew Yoder is the author or coauthor of a dozen books about radio, audio, and video, including the well-received *Auto Audio*, *Home Audio*, and *The Complete Shortwave Listener's Handbook (5th Edition)*, also published by McGraw-Hill. He has contributed articles to *Popular Electronics*, *Radio!*, *Radio World*, *New Jersey Monthly*, *Popular Communications*, and other magazines.